ENERGY RISK MANAGEMENT

Based on the proceedings of an International Review Seminar held at the International Institute for Energy and Human Ecology (Beijer Institute), Stockholm, Sweden, from September 26-29, 1978.

AN ACADEMIC PRESS FAST PUBLICATION

ENERGY RISK MANAGEMENT

Edited by

G.T. GOODMAN

*International Institute for Energy and
Human Ecology (Beijer Institute),
Royal Swedish Academy of Sciences,
Stockholm, Sweden*

and

W.D. ROWE

*Institute for Risk Analysis,
College of Business Administration,
The American University,
Washington, D.C., U.S.A.*

1979

ACADEMIC PRESS

A Subsidiary of Harcourt Brace Jovanovich, Publishers

London   New York   Toronto   Sydney   San Francisco

ACADEMIC PRESS INC. (LONDON) LTD,
24/28 Oval Road,
London NW1

United States Edition published by
ACADEMIC PRESS INC.
111 Fifth Avenue
New York, New York 10003

British Library Cataloguing in Publication Data
Energy risk management.
  1. Energy facilities – Safety measures – Congresses
  2. Energy facilities – Environmental aspects –
  Congresses   3. Risk management – Congresses
  I. Goodman, Gordon Terence.   II. Rowe, William D.
  III. Beijer institutet
  658.4'03   TJ163.15   79-42931
  ISBN 0-12-289680-7

Printed in Great Britain by
Whitstable Litho Ltd., Whitstable, Kent.

CONTRIBUTORS

AFIFI, A.A. *School of Public Health, UCLA, Los Angeles 90024, U.S.A.*

BOHMAN, H. *Skandia Insurance Company, Fack, S-103 360 Stockholm Sweden*

DENNIS, L. *NCAR, Advanced Studies Program, P.O. Box 3000, Boulder, Colorado 80307, U.S.A.*

EHRENBERG, L. *Wallenberg Laboratory, Stockholm University, S-106 91 Stockholm, Sweden.*

FISCHER, D.W. *Industriøkonomisk Institut, Breiviken 2, N-5000 Bergen, Norway.*

FISCHHOFF, B. *Decision Research, 1201 Oak Street, Eugene, Oregon 97401, U.S.A.*

HARRISS, R.C. *Mail Stop 270, NASA, Langley Research Center, Hampton, Virginia 23665, U.S.A.*

HARTE, J. *Lawrence Berkeley Laboratory, University of California, Berkeley, California 94720, U.S.A.*

HOHENEMSER, C. *Dept of Physics, 260 Science Buildings, Clark University, Worcester, Mass., 01609, U.S.A.*

KASPER, R.G. *Environmental Studies Board, National Academy of Sciences, 2101 Constitution Avenue, Washington D.C. 20418, U.S.A.*

KATES, R.W. *Dept of Geography, 2010 Academic Centre, Clark University, Worcester, Mass., 01609, U.S.A.*

LICHTENSTEIN, S. *Decision Research, A Branch of Perceptronics, 1201 Oak Street, Eugene, Oregon 97401, U.S.A.*

LINDELL, B. *National Institute of Radiation Protection, Fack, S-104 01 Stockholm, Sweden.*

LÖFROTH, G. *Wallenberg Laboratory, Stockholm University, S-106 91 Stockholm, Sweden.*

O'RIORDAN, T. *School of Environmental Sciences, University of East Anglia, Norwich NR4 7TJ, England.*

PAGE, T.   *Environmental Quality Laboratory, California Institute of Technology, Pasadena, California 91125, U.S.A.*

PEARCE, D.W.   *Dept of Political Economy, University of Aberdeen, Edward Wright Building, Dunbar Street, Old Aberdeen AB9 2TY, Scotland.*

ROWE, W.D.   *Institute for Risk Analysis, The American University, Massachusetts & Nebraska Avenues N.W., Washington, D.C. 20016, U.S.A.*

SAGAN, L.   *Electric Power Research Institute, Palo Alto, CA 94303, U.S.A.*

SLOVIC, P.   *Decision Research, 1201 Oak Street, Eugene, Oregon 97401, U.S.A.*

STAYNES, S.A.   *Monitoring and Assessment Research Centre Chelsea College, University of London, The Octagon Building, 459a Fulham Road, London SW10 0QX, England.*

THAM, C.   *Prime Minister's Office, Fack, S-103 10 Stockholm, Sweden.*

THEDÉEN, T.   *Dept of Statistics, University of Stockholm, P.O. Box 6701, S-113 85 Stockholm, Sweden.*

VEDUNG, E.   *Dept of Government, University of Uppsala, Box 514, S-751 20 Uppsala, Sweden.*

PREFACE

A strong interest in the study of risk has developed over the last ten years or so. During this time, the power of environmental hazards and natural disasters to dominate the early stages of development in human societies has been convincingly demonstrated. This perception, together with the added realization of the pervasiveness of risk from modern technology in industrialized societies, has led to a preoccupation with how risks of all kinds can be systematically avoided, reduced or otherwise controlled. Within the last few years, risk-assessment has been the subject of much study by experts concerned with human health, societal wellbeing and damage to the environment.

At the same time, the world has realized how dangerously dependent it is on an uncertain oil-resource for its energy supplies. The search for alternative solutions to the energy and development equation with drastically reduced oil is already under way. How much energy substitution can be made by using nuclear power? Or by using solar energy either as direct sun-power or indirectly via hydropower, wind, ancient biomass-derived fuels such as coal and natural gas, or modern ones like biogas and alcohol? Even as such questions are being studied, the risks to society of exploiting these alternative fuel-cycles are also being discussed.

And so to-day the recognition and estimation of risks from new energy strategies has come to play a major part in the formulation of future energy policies and the planning of new energy supplies. This is obviously a sensible step to take provided that we can recognize and quantify all the significant potential risks with enough certainty to make prudent choices. However, at present there is some anxiety that we may be misled into making bad decisions by faulty or incomplete energy-risk analysis.

A good deal of this concern is specific to the field of energy itself and arises from the practical problem that we lack adequate basic data on human mortality and morbidity, agricultural - or natural - ecosystem damage and biosphere changes, arising from current and future energy handling practices. A less tractable problem is generic to risk analysis as a whole and concerns the question of adequate methodology. Are the methods available to us for identifying, computing and comparing energy risks satisfactory and if not, can we improve them? Perhaps the most severe present-day problem, again generic to the risk process, is the "real" versus perceived risk question. In terms of energy-risk, the mismatch of perception between the technical expert trying to quantify the risks "objectively" and the public reacting, often intuitively, on quite different criteria to management policies based on these computations, has led to very real public acceptance difficulties. In a democratic society these may become electoral issues, even leading to changes of government – all for psycho-social reasons which still remain poorly understood. For this reason, several scrupulously detailed and ambitious risk studies have met with difficulty in gaining public acceptability. This problem has already stultified the debate on the nuclear energy issue in several countries.

Another difficulty has received less treatment than it merits. This is the question of how far the results of risk analysis are actually usable as management tools when implementing energy policy. The argument here is that far too little attention is being paid to the nature of the legal and institutional arrangements and management structures required to deal with energy risks. If institutional constraints or inflexible legal and management arrangements exist, slowing down or even preventing implementation of risk findings, there would seem little point in devoting long, costly studies to risk in the first place.

The International Institute for Energy and Human Ecology of the Royal Swedish Academy of Sciences (The Beijer Institute), with the support of the Swedish Energy Research and Development Commission, held an International Review Seminar in September, 1978 when these and other related topics were discussed in detail. The main purpose of this Review was to identify those areas of impact-risk analysis that would benefit from further intensive study to allow its more effective use. For this reason, the Seminar covered a wide field and the ensuing papers have been selected from the discussion programme to illustrate the problems and difficulties, and incorporated in this Volume so as to reflect current thinking about such topics. Emphasis is placed throughout the text on how usable risk-studies are as a tool in energy-policy formulation and management.

The problems themselves have been considered under five separate themes or Parts in the following pages. The first Part reviews the risk assessment approach and the opportunities open to it. It also indicates that there are real difficulties with the process, some of which may even turn out in the end to be intractable limitations on the technique itself. The remaining four Parts enlarge on this theme. The second Part discusses some of the special problems connected with risk data and their interpretation. In the third Part the topic of "objective" versus "perceived" risk is overtly dealt with. The fourth Part reviews some of the risk-assessment methods in current use and raises some questions about their deployment in the management context. The fifth and final Part is a summary of the application of some of the theoretical ideas on what constitutes an acceptable risk.

Many people helped in the preparation of the Seminar and this Volume, and deserve my warmest thanks. In particular, I am indebted to my colleague in the Beijer Institute, Dr. Lars Kristoferson and also to Dr. Torgny Schütt of the Swedish Energy Research and Development Commission. Both gave a great deal of thought and help in planning the content of the Seminar. The hundred or so experts from some seventeen different countries who attended the Seminar and contributed to our discussions also helped to shape our thinking for producing this Volume.

My co-editor Professor William D. Rowe has made the selection of papers and has taken the main responsibility for preparing short introductions to the five Parts which set each individual contribution in its overall context and give the reader guidance in selecting papers for more detailed study. He has given time and thought to this Volume so generously that it is highly appropriate that he should have the first and last word in the book in the form of the initial paper introducing risk-assessment and the final summing up on acceptable risk.

Gordon T. Goodman
Director, Beijer Institute
Stockholm

CONTENTS

PART 1
RISK MANAGEMENT: OPPORTUNITIES, DIFFICULTIES AND LIMITATIONS

INTRODUCTION

All societies, regardless of their state of development, face a whole range of environmental impacts and risks in everyday living, and have developed sophisticated mechanisms for coping with such hazards. Generally speaking, however, it has been less than a century since society has felt it could exercise some significant degree of control over the risks to which it is exposed.

In this first Part, the opening paper by William Rowe provides a short introductory review of the history and structural development of risk assessment and the present-day management approaches usually associated with it. The paper shows that we are now beginning to address the analysis and assessment of risks in a more formalized manner. The evolution of a risk assessment methodology is already under way. It offers opportunities for rational approaches to analysing and controlling risks, but also has apparent difficulties and limitations.

The remaining five papers in this Part have been assembled to display the range of opportunities which are available in human-affairs for the exercise of wise risk-management. The papers also give a foretaste of some of the difficulties encountered when trying to do so. Some of these are so intractable, at our present state of knowledge, that they may turn out in the end to be absolute limitations on the complete applicability of systematic risk-management to certain areas of policy-formulation. Each of the papers thus introduces different sets of issues – advantages and difficulties – which are dealt with more comprehensively in subsequent Parts and indeed, each of the papers could well have been included in the appropriate section later on in the book.

The second paper, by Timothy O'Riordan, takes the risk assessment categorization described in the first paper by Rowe and puts it into the setting of the total human environment, including human social-structures. In this context, O'Riordan points out the possibility that although risks may not be increasing, societal concern with risk is. As a result of this concern, efforts to combat risk in one sector may well produce new hazards in other sectors, probably affecting quite different groups of people. It therefore may not be possible to reduce the total risk-burden, a suggestion in line with mortality trends in industrialized countries.

An important point made by O'Riordan is that the very success of risk-assessment practice in identifying and quantifying hazard has caused such widespread concern in society that the additional pressure for remedial action via standard-setting and formulating codes of practice may prove too much for the risk-management process which will be overwhelmed. Are new or better institutional arrangements needed?

The third paper, by Carl Tham, provides the point of view of a political realist and practitioner. He believes that risk assessment, at least scientific risk assessment studies, can at best serve as background for political decisions that are the real focus of policy development. He rejects a "common rational ground" for the diverse risks societies bear across various sectors of activity, and maintains that excessive concern with implicit values can bog down risk assessments and subsequent political decision processes.

The next paper, by John Harte, explores the conflict between the development of technological systems, e.g. energy production, and the preservation and maintenance of the natural environment. The unique conditions at the interfaces between ecosystems and energy systems are explored in order to illuminate risk analysis as well as to add new dimensions to the study of these systems. These interfaces receive their character from the large variety of ecosystems and living processes and from the problem of non-commensurate alternatives.

The fifth paper in this Part is by David Fischer. It examines the need for reliable information and careful policy development as prerequisites for effective decision-making. Fischer attempts to identify the sources of required information and policy in the establishment of environmental standards. The methods of policy development of two different governments are described to illustrate different approaches; in particular, in an area where standards did not exist in the past.

In the final paper of this Part, Kasper provides an overview of the difficulties of handling explicit and implicit values at the policy level. He rejects at the outset the idea that any purely mechanistic and coldly analytic study can be *entirely* accurate, and states that it may even be misleading. He defines two extreme approaches: the first, to insist on "objective measure of risk" and ignore public perceptions; the second, to react only to public perceptions of risk, in order to reduce public apprehension. Either approach used alone can lead to extremes which are probably unacceptable to many.

A key problem addressed by O'Riordan is whether the risk-assessment process helps solve the problems of risk management or exacerbates them. The increased visibility of the process may bring credibility to risk-assessment institutions, but continual exposure and discussion also increases concern and anxiety.   This leads to consideration of the different approaches to risk management: the open, adversary approach of the United States, Japan and many European countries and the consensus approach of Britain and some other European countries. In the latter, risk-management is in the hands of "experts" and in some respects is an "old-boy" system. Both have their difficulties and successes. The first is slow, cumbersome and expensive, but credible; the latter is efficient, but facing an increased questioning as to whether it will continue to work. Some of the work by Slovic and Fischhoff described in Part 3 begins to challenge the capability of experts to make "good" decisions in a behavioural sense, i.e. the experts consistently seem to overrate their ability to estimate objective risks.

The risk assessment process is not rigid – only the way some people use it. Thus, if the growing power of risk analysis is beginning to overwhelm the political process, we cannot ignore the risks; but, perhaps we can make the process more effective. Both technical and political aspects must be considered in doing this, since neither bad science nor bad policy should be tolerated for long.

Another issue addressed in this Part is the conflict between short-term political expediency and the long-term impacts of technology and policy. As Tham points out, the risk analysis process must serve the political decision-maker. The process cannot by itself resolve the wide bands of uncertainty and imprecision in data or the value judgements that must substitute for precise knowledge. The political decision-maker, in establishing policy decisions, must take into account short-term political expediency. Conversely, Harte points out that problems with ecosystems are usually long-term and subtle, as their impacts are chronic rather than acute. Data are difficult to obtain, except on a localized, acute-impact basis. In this sense risk assessment may well serve to assure that the long-term impacts are explicitly considered and placed in reasonable perspective.

Both Harte and Fischer pinpoint problems of obtaining information on ecological and biological harm. Harte addresses this difficulty by stressing the need for a better understanding of environmental benefits as well as degradation.   Better monitoring systems are needed for both risk identification and for "fine tuning" the environment. Fischer on the other hand uses a case study to show how policy is often formulated with inadequate information, and raises the serious point that even if such information were made available to the various "actors", the likelihood that they would use it is questionable.

Perhaps one of the most intractable difficulties is raised by Kasper. Given that the disparity between objective and subjective risk-perception exists, propaganda and indoctrination about technical issues may be counterproductive in achieving convergence. When self-interest is apparent in such approaches, distrust and misunderstanding are exacerbated, and the rift between the objective technical experts and the public widens.

Kasper suggests a middle ground where government, the public, technical experts and developers of new technology can become involved early-on with all affected parties as participants. However, he is not at all optimistic about it ever happening.

INTRODUCTION TO RISK ASSESSMENT

William D. Rowe

Institute for Risk Analysis,
The American University,
Washington, D.C., U.S.A.

Risk is the potential for the realization of unwanted consequences from impending events. Individuals have always been exposed to risks of injury, illness or death from such events and have demonstrated a consequent safety consciousness. The last few years have shown rapid increases in both the levels of risk to which people are exposed and in the public's concern with risks, especially those resulting from new technological developments. Awareness of both immediate and long-term risks affects the manner in which society undertakes to control them.

Recognition that many risks have voluntary and involuntary components, and increased concern with regulation of risks that are imposed inequitably on parts of society, are changing the manner in which society approaches risks. Traditionally, ethical and legal questions about the appropriate extent of governmental regulation of risks have been addressed implicity rather than explicity. Thus, many problems in setting acceptable levels of risk for societal activities, especially those activities which impose risk inequitably on risk agents[1] who are not the direct beneficiaries of the risk-creating activity, are just beginning to receive attention (see Rowe, 1977a).

These changing public attitudes to risk and the newly emerging societal anxieties over risk exposure owe much to world events during the last two decades. In particular, there were growing doubts about the role of technology, generated largely by instances where its application had not been thoroughly thought out beforehand. It is useful therefore to look at such events in a historical perspective, especially in a country like the United States of America where many

---

[1] A "risk agent" is defined as the human object of risk. The term is synonymous with "risk taker", though the latter term connotes voluntary rather than involuntary risk.

especially in a country like the United States of America where many of these changes began.

Historical Perspective

The United States, which this perspective encompasses, traditionally has been a technology-orientated society. Throughout the late 19th and early 20th centuries, the concept of man conquering nature through technological skill was a dominant American cultural theme.

In the post-World War II period, a more visible structuring of this theme emerged in "the promise of science and technology" Udall, 1977) for solving all human problems. It was thought that increased investments in technological research and development and improved technological hardware would provide the solutions to such problems as energy, transportation, housing and social welfare. Nuclear energy would be so cheap that metering of homes would be unnecessary. A federal highway programme would solve the problems of transportation and personal mobility. If aerospace technology could set a man on the moon, then technology could solve man's social problems as well. Congress need only invest more in research, development, and technical programmes.

A scientific-industrial élitism accompanied this cultural theme. Those who believed in "the promise of science and technology" considered that they alone were capable of leading society to these new heights. Paternalism, characterized by a reluctance to discuss governmental issues openly, developed in many governmental institutions, such as the Atomic Energy Commission and parts of the defence and intelligence communities. Although élitist, the governmental leaders supporting technocracy were no doubt sincere in their dedication to the promise of technology. Their intent was the betterment of the human condition, their motives relatively pure, and the strength of their beliefs overwhelmed whatever weak opposition there was.

If the 1940s and 1950s were the decades of the American "promise", however, the 1960s were the years in which the promise became tarnished. The cause of this disenchantment was our increased awareness of the deleterious effects of technology, which its "promise" had failed to account for. Rachel Carson's Silent Spring (1962) aroused some interest in the side-effects of technology, especially its environmental impact. The American war in Vietnam, however, was perhaps the major cause of the changed attitude (Udall, 1977). Although American military advisers assured political leaders that this war would be won by superior technology and firepower, it became apparent that technological superiority alone could not succeed under the social and institutional conditions existing in Vietnam.[2] The deleterious side effects of the technological weapons

---

[2]Perhaps the major causes of this failure were a corrupt South Vietnamese government, a dedicated and indigenous communist

used in Vietnam were often more evident than their successful application. The effort to remove trees and scrub, used as military camouflage cover, with "Herbicide Orange", [3] for example, was never particularly successful, yet the spreading of Herbicide Orange on civilian food crops and into water supplies resulted in birth defects (National Research Council 1974).

As a result of this disillusionment, America's youth – the Americans most affected by this war – challenged the military-industrial complex directly for the first time. Others became enamoured with the life-style of the anti-war movement, and the so-called "counterculture" developed. This group considered technology immoral and criticised institutions supporting technology. In the past decade, however, the counterculture has matured beyond anti-war violence and demonstrations, and has begun to effect change peacefully through the legislative process.

The environmental movement developed concurrently with the counterculture; both represented political forces opposing the uncontrolled use of technology. This anti-technology movement has not been without impact. Congress enacted the National Environmental Policy Act of 1969 (NEPA), requiring environmental analysis in technological and political decision-making. Nuclear energy has become the focal point for the anti-technology movement, no doubt because of its high technology content, its relationship to atomic weapons, its cancer-causing potential, and the scientific elitism of the atomic establishment (Lapp, 1975). Perhaps, too, this antagonism developed because scientists have learned more about nuclear energy and its effects than about other energy-producing technologies. If so, the political reaction to nuclear proliferation is only a precursor of the critical public scrutiny that all new technological systems can expect to meet.

The exact significance of the anti-technology movement is difficult to assess because those who oppose the "promise of technology" are different from those who oppose technology itself. The difference is distinct. The former challenge government and industry to assure that technology is used in the best interests of society; they seek to avoid imposition of unwanted or undesirable technology, but recognize the need for a technological growth in modern industrial society. The latter oppose the development of any technology,

---

effort, and an American public disillusioned by "kill ratios", atrocities, and the mounting casualty lists of reluctant American draftees.

[3] The defoliant consisted of a 50-50 mixture of the n-butyl esters of 2, 4-D (2,4-dichlorophenoxyacetic acid) and 2, 4, 5-T (2, 4, 5-trichlorophenoxyacetic acid). This mixture contained teratogenic impurities.

regardless of its potential benefits.[4]

Starting with enactment of NEPA, the search for institutional solutions to the risks of technology characterizes the 1970s. The Environmental Protection Agency (EPA) was formed in 1970, followed shortly by the Occupational Safety and Health Administration and local consumer protection authorities. The Congressional Office of Technology Assessment was established in 1972 to evaluate the long-term effects of new and existing technologies. These institutions, formed to counter the inequities imposed by technology on parts of society, began to systematically address the long-term impacts of technological developments rather than simply their short-term effects.[5]

Institutional approaches to technology assessment can operate successfully only if institutional decisions are accepted by the public. Credibility requires recognition that governmental decisions must be made in a visible and open manner that encourages participation by all parties, including the concerned public. The process by which such decisions are made has in this sense become as important to society as the decision itself (Lovins, 1977; Merrill, 1977).

The next decade may see successful implementation of these institutional approaches to technology assessment. Much effort has been spent determining how best to use new institutional requirements such as the environmental impact statements required under NEPA. Scrutiny of such requirements indicates that the institutional approach can be effective, and massive paperwork reduced significantly. Nonetheless, a variety of problems must be solved before effective institutional approaches can be established to assess societal risks.

---

[4] Anti-technologists desiring a return to a pastoral existence are wishful thinkers; our planet cannot support 6 billion or more people without specialization, effective use of resources, and a technological base. The type of technology employed is open to revision, however, and the "distributed" technology suggested by E.F. Schumacher (1973) for developing nations is a thought-provoking alternative to present systems.

[5] Both the executive and legislative branches of the United States government conduct their decision-making functions prospectively, in a political arena and in an atmosphere of crisis management. The judicial branch, by contrast, only operates retrospectively after damage is done or anticipated damage proven. Responsibility for the long-term consequences of technology must be shared by all three branches. NEPA, for example, gives both the executive and the judiciary means to handle long-term problems, while the Office of Technology Assessment helps Congress address the long-term environmental impacts of new or proposed programmes.

The Process of Risk Assessment

The process by which risks are assessed increases in importance as society becomes more cognizant of risk, particularly risk that is inequitably imposed on certain individuals by the technological activities of their society. The term "risk assessment" is used here to describe the total process of risk analysis, which embraces both the determination of levels of risk and the social evaluation of risks. Risk determination consists of both identifying risks and estimating the likelihood and magnitude of their occurrence. Risk evaluation measures both risk acceptance, or the acceptable levels of societal risk, and risk aversion, or methods of avoiding risk, as alternatives to involuntarily imposed risks. The relationship among the various aspects of risk assessment is illustrated in Figure 1.

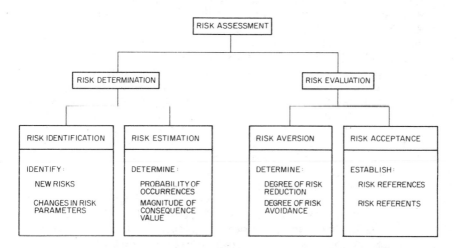

Fig. 1 The components of risk assessment

The major focus of the ensuing papers in this book is risk evaluation, especially risk acceptance. To provide the reader with a balanced background, risk determination will be emphasized.

Risk identification is of particular interest because modern technology has created many new risks that are either irreversible or of global proportions (e.g. release of long-lived radionuclides, such as plutonium and carbon-14, to the environment). Though risk identification is an endless task, little effort has until now been made to identify even major types of risk prior to their imposition.

Risk Determination: Identification

If risk is defined as a potential threat to man, his environment, and his well-being, the manner in which new risks are perceived and identified is a major consideration in coping with them. Estimating changes in the level of man-made and natural hazards[6] requires both measurements of changes in levels of risk and evaluation of the perception of such risks. The two are mutually dependent; the former process identifies new rises, while the latter determines social and cultural response to such risks. Changes in the levels of risk are identifiable in three circumstances: when a new risk is created; when the magnitude of an existing risk changes; and when perception of an existing risk changes. All three may occur simultaneously.

A new risk is a risk that did not previously exist, not a new identification of existing risk. New risks are almost always created by man, usually as a result of a new technology against which nature often has no natural defence.

A new perception of existing risks may arise because the hazard has been identified, because the magnitude of the hazard changes suddenly, or because a slow change in the hazard's magnitude crosses some threshold of societal concern. Alternatively, public perception of existing risks may change whether or not the level of risk is changing in an objective sense. This change may occur when a more dominant risk is eliminated or reduced, such as when the reduction of infectious diseases transfers public concern to chronic diseases. Or, public concern with risk may be stimulated by communications-media, or by the threat that certain hazards pose to particular individuals or groups rather than statistically to the population at large. Individuals satisfied with the *status quo* may view the hazard as a threat to their present condition.

Risk identification efforts in the United States currently focus on screening chemicals for toxicity and carcinogenesis and identifying technological threats to geochemical systems. In the former area, the newly enacted Toxic Substances Control Act (1977) authorizes the EPA to obtain from industry data on the production of chemicals, their effects on the health of users, and other matters relating to chemical substances and mixtures. The EPA may regulate when necessary the manufacture processing, distribution, use and disposal of particular chemical substances or mixtures.[7] In the latter area, studies are underway to identify technological threats to geochemical systems, such as the threat posed to the ozone layer of the earth's atmosphere by fluorcarbon gases and high altitude jet travel (Molina, 1976), by thermal and carbon dioxide loading to the earth's climate,

---

[6] The terms "hazard" and "risk" may be used interchangeably, though the term "hazard" connotes the present existence, and "risk" the potential existence, or "threat" of some unwanted consequence.

[7] Pesticides, tobacco, nuclear material, firearms and ammunition, food, food additives, drugs, and cosmetics are exempted from the Act because they are regulated under other statutes.

and by new toxic materials to the unprotected environment. Several inter-governmental agencies (World Meteorological Organization, World Health Organization, U.N. Environment Programme) are conducting studies (see Lane, 1977).

Risk Determination: Estimation

The process of risk estimation has five steps: identifying the cause of risk; measuring its effects; determining risk exposure; defining the consequences of exposure; and valuing the consequences of exposure. These five steps are illustrated in Figure 2.

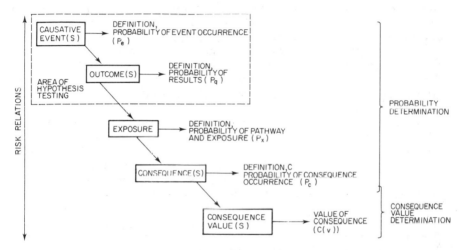

Fig. 2 Process of risk estimation

The first step in risk estimation is identifying causative events, or events that create a probability of risk occurrence. Each causative event may lead to several possible outcomes. In the second step, these outcomes are defined and their relative probability determined. The causative event and outcome do not of themselves constitute risk until the exposure of humans, institutions, and the natural environment is considered. The probabilistic relationship between events and their outcomes can be measured through empirical experiments, statistical design of experiments, and hypothesis testing. If such experiments require exposing humans or the environment to risk, however, then the risk of conducting the experiments must also be evaluated.

The third step in risk estimation is defining exposure pathways, the means by which risks are transmitted. Exposure pathways can be defined explicitly; for example, a pesticide may reach man through food, water, and aerosol pathways after application to a food crop. The probability of various exposure pathways and of resultant exposures is also determined in this step. The fourth step is defining the possible consequences of risk exposure and determining for each risk the probability that consequences will occur. The final step in risk estimation considers the value placed by affected individuals on the consequences of risk exposure. Some individuals, for example, may be unconcerned about consuming trace amounts of pesticides in food because they consider the effects remote; others may be sufficiently concerned to eat only organically grown foods.

The probability of risk occurence and the value placed on risk consequences by affected individuals determines the public's response to risks. Thus, the process of risk estimation requires two basic determinations: a consequence probability determination and a consequence value determination. The steps comprising each determination are bracketed on the right side of Figure 2; the determinations overlap at step 4 consequences.

The estimated risk, however, cannot be fully equated with actual risk because probability and consequence estimates may be inexact. The probability that a consequence will occur may be determined by direct observation of repeated trials of a causative event. When the number of trials is large, the estimate of probability is properly considered objective, because it represents an empirical estimation. On the other hand, if probability estimates are made on the basis of one or a few trials or by conjecture, then the probability estimate is subjective. These definitions comport with classical definitions of objective and subjective probability.[8] Between these extremes lies another estimate, here termed "synthesized probability," for which the probability of a consequence is not measured directly, but rather is extrapolated from the objective probabilities of causative systems that are expected to behave similarly. For example, the Nuclear Regulatory Commission's study on reactor safety (Rasmussen, 1975) uses an estimate of probabilities computed from tests on parts of the reactor system and synthesized into a total model.

A similar range of certainties exists in calculating consequence value. An objective estimate of consequence value consists of a consequence that is directly observable and measurable, and a consequence value that is expressed explicitly. For example, the accounting value of the pay-offs of a gambling establishment represents an objective consequence value. Subjective consequence value, on the other hand, occurs when the consequence value to a

---

[8] Subjective probability is a number between 0 and 1 assigned to an event, based on personal views as to whether the event will occur. Objective probability is a number between 0 and 1 assigned to an event, based on a history of trials of similar events, estimating whether the event will occur (Collocott and Dobson, 1974). Concerning subjective estimates, the Bayesian approach to conditional probability and the use of a *priori* information is useful (See Schlaifer 1969).

particular risk agent depends completely on the risk agent's personal value system and situation. Between these two extremes is a value estimate, here termed "observable consequence value," in which the behavioural response of groups in society to objective or subjective risk consequences is ascertained and measured by studying actual behaviour.

A relationship between the nature of probability and the nature of consequence, both of which range from the objective to the subjective, is illustrated in Figure 3. Probability is diagrammed

PROBABILITY	CONSEQUENCE / OBJECTIVE CONSEQUENCE EVENT DESCRIPTION WHICH IS DIRECTLY OBSERVABLE AND MEASURABLE	OBSERVABLE CONSEQUENCE MEASURED BEHAVIOURAL RESPONSE OF GROUPS TO OBJECTIVE OR SUB-JECTIVE CONSEQUENCES	SUBJECTIVE CONSEQUENCE VALUE OF A CONSEQUENCE TO A PARTICULAR RISK AGENT
OBJECTIVE PROBABILITY MEASURED BY REPEATED TRIALS	OBJECTIVE RISK	MODELLED* RISK (VALUATION)	SUBJECTIVE RISK (VALUATION)
SYNTHESIZED PROBABILITY MODELLED FROM SIMILAR OBJECTIVE PROBABILISTIC SYSTEMS, BUT NOT MEASURED.	MODELLED* RISK (ESTIMATE)	MODELLED* RISK	SUBJECTIVE RISK
SUBJECTIVE PROBABILITY ESTIMATED FROM FEW TRIALS OR THROUGH CONJECTURE.	SUBJECTIVE RISK (ESTIMATE)	SUBJECTIVE RISK (ESTIMATE)	SUBJECTIVE RISK

* TO THE EXTENT THAT THE MODEL CORRESPONDS TO REALITY, THE MODELLED RISK APPROACHES OBJECTIVE RISK.

Fig. 3   The subjective and objective nature of risk measurement

vertically and consequence horizontally. The combination of probability and consequence defines risk. Thus, the combination of objective probability and objective consequence defines objective risk. Most scientific studies of risk have concentrated on objective risk because such risks are easiest to define and measure. As the diagram moves toward synthesized probability or observable consequences, it defines an area termed "modelled risk." Such risks are neither directly observable nor objective, and the usefulness of computing modelled risk depends on the degree to which the model corresponds to reality. Such risks may be modelled by using synthesized probability, or observable consequences, or both. All other risks are subjective because they are computed on the basis of subjective estimates, subjective valuations, or both.

The science of risk estimation has traditionally used scientific experiments and empirical measurements to compute objective risk. Recently, the concept of synthesized probability has developed extensively, just as measurement of observable consequences has gained major importance in the behavioural sciences. By contrast, societal decision-making relies on subjective risk estimates; in practice, emotional and political considerations are more compelling than objective scientific knowledge. Part 3 of this book addresses this dichotomy in depth.

Risk Evaluation

One cannot evaluate risk without considering human responses. Man is by nature risk aversive, tending to avoid acts or situations whose consequences may threaten his health and safety. Although man may choose to take or accept risks to achieve specific benefits when such choice is self-imposed and under his direct control, he will avoid non-beneficial risks imposed on the human population by man or nature. That human society is more concerned with the adverse consequences than the benefits of risk is illustrated by society's news preferences. Reporting of disasters, crimes and other disagreeable news far outweighs descriptions of positive human achievements and other beneficial events.

The risk aversive nature of society, together with increasing awareness of new risks created by new technology, has focused attention on technological risk. The growth of public awareness and concern for risk is probably irreversible because the knowledge required for technology assessment and risk identification is now available to all levels of society. Consideration of societal risk has become an accepted part of technological decisions, and increased regulatory attention is focused on risk assessment. A methodological approach that assures a reasonable perspective in evaluating risk is necessary if a government's regulatory apparatus is to work responsibly and visibly.

The degree to which risks are voluntary affects both their public acceptability and the degree to which governmental regulation is appropriate. Purely voluntary risks, which result only in direct gains and losses to the risktaker,[9] need not and probably should not be regulated by government. Unfortunately, these voluntary risks are rare. Most voluntary risks impose some indirect risk on others who do not share the concomitant benefits. For example, the act of suicide affects not only the suicide victim, but his survivors, his insurance company, his creditors and his society. The benefit of the suicide to the individual who chooses to kill himself is subjective, situational and emotional. Moreover, both religion and government regulate the act of suicide in such a way as to make it as unattractive as possible. Thus, government and other societal institutions may regulate voluntary risks that incur some indirect losses, as well as involuntary risks that incur purely involuntary losses.

---

[9] The author prefers the terms "gain" and "loss" to "benefit" and "cost," because the latter terms connote more subjective measurement.

When indirect losses associated with a voluntary risk are imposed on significant numbers of the population or on identifiable recipients, the risk may become unacceptable, and regulatory action to ameliorate the risk-inequity becomes necessary. A process for determining levels of risk acceptance must consider the effects of these involuntary risks. In the final analysis, the purpose of a risk acceptance process is not to balance direct gains and losses of risk taking, but to reduce inequities in indirect gains and losses. Risk acceptance results only when direct and indirect losses are favourably distributed. Thus, a risk acceptance process is neither a cost-benefit analysis nor a substitute for such analysis. Its purpose, recognizing that some levels of risk always exist, is to determine when risks imposed on segments of a society are low enough to be acceptable. The weighing of indirect gains against imposed risks is a component of risk acceptance; higher levels of risks may be acceptable by society if indirect benefits are received along with losses. For this reason, total societal equity in risk imposition is rarely achieved in practice.

The institutionalization of risk assessment in governmental decision-making requires a variety of administrative determinations. Such determinations include identifying individuals or agencies that should evaluate risks for society; determining the type of regulatory agencies that can evaluate risk credibly and defining their respective roles; and devising credible methods by which the value judgments necessary to the risk evaluation process can be made.

The value judgments required in risk evaluation fall into three classes: technical, societal, and managerial. Where hard technical information or data is absent, or when information gathering is too costly or time-consuming, technical value judgements concerning risk are made by technical experts who may disagree on the consequences of particular activities.[10] Scientific bodies, such as the National Academy of Sciences, therefore usually make technical judgments by consensus, as in the BEIR Report (National Research Council, 1972; Lovins, 1977) on low-level ionizing radiation.

A societal value judgement of risk weighs the benefits and costs of risks and attempts to minimize inequities in the cost-benefit balance. Scientists and technical experts have no more expertise in

[10] In the author's opinion (Rowe, 1977b), a so-called "science court" could not substitute for such scientific judgment. "An underlying fallacy (in the concept of a science court) is the notion that science, unlike most fields of human endeavor, reaches solid, indisputable conclusions that are in essence, value free. In reality, science at the frontiers of knowledge, which is where most public policy disputes arise, seldom even pretends to be the discovery of absolute truths. The things that scientists can agree about often are so few as to contribute very little to policy making."

this area than other well-informed members of society. Interested individuals, societal groups, or governmental agencies themselves may express value judgements on behalf of the public in governmental proceedings. By contrast, the managerial value judgement of risk interprets and modifies the societal value judgment in order to implement and enforce public opinion concerning acceptable risk. The state implementation plans mandated by the Clean Air Act, for example, require managerial value judgments and mandate the participation of lawyers, law enforcement personnel, scientists, engineers, and government administrators, as well as technicians. All three types of value judgements – technical, societal and managerial – must be made and weighed in governmental decision-making concerning new technologies.

References

Carson, R. (1962). "Silent Spring". Houghton Mifflin, Boston, Massachusetts.

Collocott, T.C. and Dobson, A.B., eds. (1974). "Dictionary of Science and Technology", 720-726 W. and R. Chambers, Edinburgh.

Lane, (1977). A problem of hot air. *Newsday*, May 28, p.A4.

Lapp, R. (1975). "The Nuclear Controversy". Fact Systems, Greenwich, Connecticut.

Lovins, A.B. (1977). Cost-risk benefit assessments in energy policy. *George Washington Law Review* **45**, 911-943.

Merrill, R.D. (1977). Risk-benefit decision-making by the food and drug administration. *George Washington Law Review* **45**, 994-1012.

Molina, M. (1976). The validity of the ozone depletion theory. *Proceedings of the National Bureau of Standards Conference on Photochemistry.* U.S. Government Printing Office, Washington, D.C.

National Environmental Policy Act of 1969, 42 U.S.C. 4321-4347 (1970).

National Research Council (1972). "Biological Effects of Ionizing Radiation". National Academy of Sciences, Washington, D.C.

National Research Council (1974). "The Effects of Herbicides in South Vietnam". National Academy of Sciences, Washington, D.C.

Rasmussen, N., ed. (1975). "The Reactor Safety Study". Report WASH 1400. U.S. Nuclear Regulatory Commission.

Rowe, W.D. (1977a). Governmental regulation of societal risk *George Washington Law Review*, **45**, 944-968.

Rowe, W.D. (1977b). The Science Court. *Wall Street Journal*, January 21, p12 (also reprinted in Rowe, W.D. (1977) "An Anatomy of Risk", Wiley, New York).

Schlaifer, R. (1969). "Analysis of Decisions Under Uncertainty". McGraw Hill, New York.

Schumacher, E.F. (1973). "Small is Beautiful". Blond and Briggs, London.

Toxic Substances Control Act, 15 U.S.C.A. 2601-2629. (West Supp., 1977).

Udall, S. (1977). The failed American dream. *Washington Post* June 12, B1.

ENVIRONMENTAL IMPACT ANALYSIS AND RISK ASSESSMENT IN A MANAGEMENT PERSPECTIVE

Timothy O'Riordan

*University of East Anglia,
Norwich, U.K.*

The Scope of Environmental Risk Management

Coping with environmental risk has been an intrinsic aspect of life for all societies throughout history. The process of gaining a living necessarily involves risks or adverse environmental consequences, since the natural environment is not totally benign, and man's understanding of its working mechanisms is far from complete. Thus in any search for the environmental goods regarded as necessary to improve social wellbeing (food, energy, materials), there will be associated environmental costs, some of which will be observable, some calculable and some unknown. This is the case for all societies, regardless of their state of "scientific" knowledge or technological development, so it is worth observing that all social groups have evolved quite sophisticated coping mechanisms in order to live with environmental risk (Kates, 1977, 1978; Burton *et al.*, 1978).

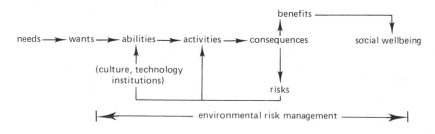

Fig. 1   Environmental risk in the context of social gain.

Figure 1 describes this relationship in simple, diagrammatic terms. It shows that environmental risks are the unintended but inevitable by-products of resource exploitation which itself is the outcome of the collective striving for a better life. The diagram also implies that the terms "environmental impact" and "environmental risk" are broadly one and the same thing. Conventionally speaking, impact is the outcome and risk the probability of its occurrence, but nowadays many people prefer the collective phrase "environmental risk" to include both an adverse event and its likelihood (Council for Science and Society, 1977; Burton and Whyte, 1980). Thus the term "environmental impact analysis" (EIA), commonly employed in the early seventies following its emergence in the US National Environmental Policy Act, is now being incorporated in the more inclusive concept of "environmental risk management". Likewise, the more specific term "risk assessment" (RA), or the calculation of the probabilities of hazardous consequences of certain courses of action, is also encompassed within the risk management concept. The advantage of this fusion of terms lies in the linkage between analysis and control, for the notion of risk management includes the institutions and regulations designed to cope with risk as well as the methods for calculating it. EIA, on the one hand, is regarded principally as an analytical accounting device, while on the other, risk assessment is seen more as a modelling and balancing mechanism. These distinctions will be elaborated in the second section of this paper.

Figure 1 also indicates that environmental risk management encompasses the links and associated learning loops between social and technological processes (abilities), the application of these processes (activities) and the effect on social wellbeing in terms of net or residual gains (sometimes referred to as risk-benefit analysis). To date, it has not normally included investigation of whether the particular demands made by a society on its environment are legitimate or could be met by other means. Such almost metaphysical debate is however important and quite legitimate, so it is worth observing that institutions established to deal with environmental risk have still not responded adequately to demands to widen the scope of discussion. Thus it is suggested (Otway *et al.*, 1978) that anti-nuclear protesters are as anxious about the kind of society that will require nuclear power as they are about the actual radioactive risks associated with nuclear power production. It is doubtful that risk assessment techniques, in their purely analytical sense, will ever be able to incorporate the dimensions of such anxieties, for these feelings involve social and political judgements that lie beyond the realm of risk calculus. Whether risk management institutions can ever satisfactorily merge technical analysis and political judgement forms a major theme of the third section of this paper.

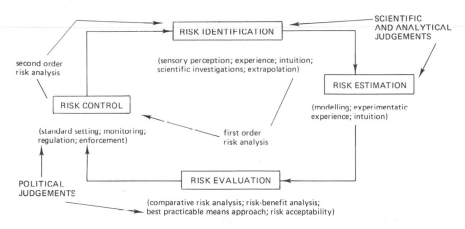

Fig. 2 Environmental risk management functions.

The task of environmental risk management identified in Figure 1 is described in more detail in Figure 2. The task is really a search for the safest route between social benefit and social loss, and is often regarded as a balancing process in which different combinations of risks are compared with various beneficial outcomes.

In general terms, environmental risk management consists of four components which in any given instance do not necessarily follow each other in a logical sequence. *Risk identification* is the activity of recognizing that a certain risk exists. This is obviously a very important function but one that is dependent on knowledge, research findings, social experience and intuition.

One interesting feature of the recent obsession with risk analysis is that it has uncovered all kinds of potentially adverse effects heretofore unsuspected. This is an outcome of improved measurement capability as much as it is a product of better scientific understanding. Concentrations of substances in parts per trillion require a degree of measurement accuracy that was unimaginable twenty years ago, but now place an enormous burden on monitoring facilities. The obvious danger here is that of exaggerating the scale and magnitude of risks beyond reasonable compass for investigation, and thus making all the other components in risk management much more complicated and time consuming. For example, the full risk assessment of all toxic products required to be examined under the US Toxic Substances Act (1976) will take at least ten years and cost millions of dollars. Major chemical corporations are seriously beginning to question whether they can produce a commercially viable new product given the time and expense necessary to evaluate the potential risks. (Risk investigation alone now takes 25-30 percent of all research and development funds involved in pesticide development according to estimates made by Dow Chemicals in the US and I.C.I. in the UK).

Needless to say, it is unlikely that even with the most sophisticated studies all the relevant risks will be identified, so it should be noted that one important feature of the risk identification phase is the search for "second order risks" – namely the risks associated with the technologies and techniques developed to reduce the first-order perceived risks. An obvious example is the switch from underground coal production to opencast mining. Each technique contains a different package of risks for different groups of people and different surrounding environments. It is questionable whether these two "risk ensembles" can ever sensibly be compared in purely analytical terms, that is, without resort to wider political judgements. Nevertheless, the fact that "risk chains" need to be considered links the four components of risk management in a cyclical manner as portrayed in Figure 2.

Risk estimation involves the modelling of risk causes and pathways, the prediction of likely consequences and their distribution in space and time. This is a much more difficult task than is often imagined. Though enormous research funds are spent on modelling ecological consequences, the methodologies and the funding for investigating the social impact of, say, major alternative energy proposals, are still in a very primitive state. One wonders whether adequate research techniques can ever be developed and proper levels of funding can ever be provided for the social aspects of risk estimation, but one can be sure that any effort in this area will be enormously time consuming, fearfully expensive and continually subject to analytical criticism (the difficulties are described by McEvoy and Dietz, 1977).

Risk evaluation is really the central task of risk management for it involves not only the social judgement of risks as identified and estimated, but the balancing of such risks against perceived and/or estimated social gains. The bulk of this paper will be devoted to an appraisal of the social and political implications of the risk evaluation function which is emerging as such a central feature of modern resource management.

Risk control is the implementive phase of risk management, the point where executive action is the form of regulation, monitoring, enforcement and re-evaluation take place. Ideally, these tasks are undertaken following an extended process of research, consultation and judgement, but in many individual instances those logical steps are short-circuited due to lack of time, political expediency, insufficient funding or adherence to established procedures regardless of their unsuitability in the light of new circumstances. Probably the primary short circuiting cause is political expediency. In the past ten years there have been some famous examples of products removed from the public domain long before all the necessary scientific investigations were completed. There are also many examples of standards that are excessively strict (for instance the requirement by the European Commission that certain toxic substances should be removed to the point of zero detection) which suit symbolic political interests but which may be quite impracticable to enforce. These illustrations serve to indicate the differences of perception that divide scientific evaluation from political judgement.

It is worth commenting that advances in the first two phases of environmental risk management (namely risk identification and estimation) have not yet been absorbed in changes to the latter two phases, and that this lag effect is producing some tensions within the overall function of risk management. As mentioned earlier, the implications of this lag effect on the evaluation and control functions will be discussed in the sections that follow.

One final introductory point should first be made. This concerns the reasons why risk management has suddenly leapt into prominence in environmental policy making. There are probably three major reasons for this:

1) Since the Second World War, advanced technological man has invented a variety of substances and processes that are potentially dangerous to very large populations of humans, animals and plants. In many instances these hazards place an involuntary burden on people who have limited means to avoid them or to influence the officials and institutions that seek to reduce such risks on their behalf. Because of improvements in measuring and experimental techniques, the identification and estimation of these risks have become highly specialized procedures which have led to interim but controversial findings and caused great public alarm. There is no doubt that this anxiety is fanned by the seeming impotence of informed lay people, frustrated at their inability to change the course of events.

2) The very products and technologies which are causing such alarm in developed countries are now being conveyed to developing countries which have neither the experience nor the management institutions to cope with them adequately. In some cases the imported technologies are highly inappropriate given the nature of economic activity and indigenous knowledge, but they are being adopted without full understanding of the wider social and ecological consequences. This is particularly the case with chemical pesticides, which are already proving a hazard to users and to benign flora and fauna. It is also evident in such areas as civilian nuclear technology and the adoption of so called "appropriate" energy production systems (based on the sun, wind and biogas) any of which is bound to have important social and economic ramifications which may not (or cannot be) fully calculated in advance. It is therefore all the more necessary to upgrade risk management techniques generally, and to ensure that these are fully adapted to suit the needs of different

societies. (Here is where the application of risk management ought to cover the whole spectrum portrayed in Figure 1).

3) There is now some justification for the assertion that it may no longer be possible to reduce *total* environmental risk. As pointed out earlier, efforts to combat risk in some sectors may well produce new hazards in other sectors, probably affecting different groups of people. So we may be reaching some kind of "environmental risk steady state" in which extremely costly efforts to reduce risk merely stop the whole penumbra of risk from getting worse. If this state of affairs is actually the case, then great emphasis will be placed on devices to improve comparative risk-benefit assessment procedures, though it is doubtful that any comprehensive methodology will ever be satisfactory. Comparative risk analysis and risk-benefit analysis must eventually enter the realm of politics. It is perhaps ironic to note that the very success of risk identification and estimation functions has contributed to the "risk steady state" belief, since our state of ignorance is now being reduced.

The consequences of all this activity in risk identification and estimation for the two decision-making functions in risk management, namely evaluation and reduction, are profound and somewhat disturbing. First of all, the risk management process is controlled so much by these technical phases that it is regarded more as a mechanism for scientific investigation and rational control than as a properly established political process by which a range of societal values are fully considered. Secondly, as a consequence of this, the whole risk management function has become excessively specialized and technical to the point where many ordinary people are hopelessly confused, and politicians have become alienated from the processes of scientific judgement that lie behind the advice upon which they depend. So important matters of substance can become swept aside in bitter wrangles between prominent scientists anxious to protect their egos and their reputations. There are many aspects of the nuclear dispute that have fallen foul of this problem (e.g. the "hot lung" debate, the argument over acceptable amounts of low level radiation exposure over a lifetime, and disputes over the "pathways" by which radioactive emissions may affect animals and mankind). And thirdly, the outcome of this is that the customary processes by which conventional political decisions are made (e.g. consultation with key lobbies; information availability and comprehensibility; public participation; accountability of regulatory officials and their political masters and procedures for arbitration such as parliamentary debates, the courts, public inquiries and adjudicatory tribunals) are all proving inadequate in the face of this growing sense of rational scientific control and highly technical sophistication.

It is evident, for example, that the views of key lobby groups may be bypassed (e.g. farmers in the case of pesticides; the energy utility companies in the case of nuclear radiation; the chemical companies in the case of toxic products) in favour of influential scientific advice over regulatory standards and tough requirements for monitoring. But in the actual conduct of business, such restrictions may be ignored or at least reinterpreted simply because they may not have taken into account the feelings and customary practices of such groups. Indeed, the very rigid structures inherent in the risk control function may impede its effectiveness. It is equally evident that an excessive amount of discussion during the task of evaluation (which properly is a political matter) can be taken up with detailed scientific controversy regarding the merits of one particular set of findings over another. In short, the growing power of risk analysis techniques is presently influencing established political processes in ways for which the latter are not adequately prepared.

Before discussing how risk management institutions might have to adapt to these new scientific demands, it is worth looking briefly at the advantages and disadvantages of the twin techniques of environmental impact analysis (EIA) and risk assessment (RA) in terms of their influence on policymakers.

Environmental Impact Analysis

Environmental impact analysis emerged legally in Section 102 (C) (?) of the US National Environmental Policy Act, which became law on January 1st, 1970. Since there have been many studies of its legal evolution and interpretation (Anderson, 1973; Dreyfus and Ingram, 1976; Atherton, 1977; Fairfax, 1978), no further analysis of its legislative importance will be provided here. But it is worth recording that EIA was originally regarded as a descriptive accounting device to complement the controversial use of cost-benefit analysis and that its aim was to make executive agencies aware of the need to consider the wider ecological effects of their actions (not readily calculable by cost-benefit analysis), and of the value of consulting with other government departments before major resource development schemes were put into effect (see Caldwell, 1977). It is also worth commenting that, at the time the legislation was passed, few Congressmen and even fewer agency officials had any idea of the enormous procedural demands this clause was to make, and certainly very few people regarded the EIA mandate as an opportunity to reappraise the whole logic of resource development and environmental manipulation.

Over the years environmental impact analysis has grown up into a reasonably systematic technique whose findings are publicly available for inspection and which has aided public officials and politicians in understanding some of the wider and longer term implications of certain projects and policies. It has encouraged a great improvement in investigatory techniques, enables interagency communication to operate more smoothly, and is providing a variety of consultants (some *bona fide*, others charlatans) with a comfortable living. It is now so institutionalized in the US agency apparatus as to be regarded as a bothersome but necessary preliminary for the justification of any particular resource development project (which would include all major energy projects).

But few students of EIA believe that it has fundamentally changed predominant values about resource exploitation and environmental disruption.   EIA is an administrative device which is applied to particular proposals; it is not widely employed as a mechanism to judge the merits of a particlar policy or course of action. In other words, in many cases certain commitments are made and budgets allocated before the substantive aspects of the EIA procedure are set in motion, thereby weakening its influence over the possibilities of considering quite different approaches to solving a particular resource management problem. In the area of energy policy making, for example, EIA is not widely used to consider the merits of strategies to reduce demand over strategies to increase supply, let alone to help in the establishment of priorities for budgetary allocation within each of these two strategies. EIA tends to enter once major policy options have been established – that is, it is undertaken on particular proposals (the fast breeder reactor) or specific schemes (an actual energy generating plant) over which crucial philosophical, political and budgetary commitments have already been made.

Given this context, it is hardly surprising to learn that EIA is now regarded in some quarters as rather a cumbersome procedure that is costly, time-consuming and decision-extending without adequately satisfying its primary purpose of protecting the environment (Dreyfus and Ingram, 1976; Fairfax, 1978). While it may have helped to modify the design characteristics of very large scale developments (e.g. the trans-Alaskan pipeline) rigorous application of EIA has rarely stopped a major project in its tracks. One notable exception here is the case of the fast breeder reactor whose present construction moratorium was in part the result of an adverse environmental as well as economic appraisal.

In sum, the evidence on the merits of EIA is ambiguous and contradictory. Agency officials and environmental consultants praise its administrative and political significance, pointing to important changes in agency thinking with respect to environmental intervention. (The US Corps of Engineers and Forest Service are frequently cited in this regard). Academics and environmental lobbyists, on the other hand, are more critical in their assessment, arguing that it has not fundamentally altered agency responsiveness, that a new form of verbal secrecy (to circumvent the Freedom of Information Act) is being developed, and that the legal status of citzens to review the merits of agency decisions or to be guaranteed some respectable level of environmental quality has not substantially been altered.

Despite these criticisms it can be said that EIA has permitted greater public access to agency thinking, it has improved the taxonometric display of various kinds of outcomes and it has stopped the more hare-brained resource development proposals. But in its acceptance and institutionalization it has tended to become standardized and elaborated to the point of being an administrative nightmare. In the long run it needs to be refurbished and extended into the policy evaluation arena if its merits are to be fully realised. This will prove to be a very difficult and unpopular task.

Risk Assessment

As mentioned earlier, risk assessment is still an emerging technique, so it is premature yet to evaluate its long-run contribution to environmental decision-making. As is the case with EIA, its principal advantages lie in its logical ordering of causes, pathways and outcomes, each of which can be detailed to a considerable extent. In addition, RA is making important demands on mathematical probability theory, and on the theories and techniques in social psychology, to enable researchers and decisionmakers to understand how concepts about risks and benefits are formed among various groups and subsequently acted upon. Also, it is bringing about improvements in the technologies of risk reduction – as anyone who has worked in the safety aspects of nuclear power production will confirm.

Already there are substantial achievements in each of these three important areas even though there are still many improvements to be made. But at least it is now possible to improve our understanding of the difficult concept of probability and to apply it to the way in which ordinary people conceptualize unpredictable events. The advantages here are enormous, for the fruits of risk assessment research are beginning to enable people who are risk acceptors to cope more knowingly with various kinds of technological risk, to recognize the merits and disadvantages of different technologies and behaviours aimed at reducing risk, and thus to communicate more sensibly with politicians who must decide on various appropriate courses of action. The research here is still at a very early stage of development, so it is still far too soon to tell whether the results of this work will genuinely enable politicians to make better decisions in the public interest. But the preliminary findings, though somewhat ambiguous, do indicate that ordinary people can distinguish between different kinds of risk, that their interpretations are altered if they are given comprehensible scientific evidence, and that they can at least recognize a relationship between risk and benefit. Should these findings be widely substantiated the educational requirements that lie behind risk evaluation could be formidable, but plausibly justifiable.

Nevertheless, there are a number of disadvantages inherent in risk assessment which still have to be overcome. The first is the everpresent danger of scientific rationality overwhelming political rationality. These two concepts of rationality are widely different but tricky to define. Scientific rationality is based on the tenets of empiricism and logical positivism, whereby laws are deduced from experimental evidence and experiments devised to test laws. Pure scientific rationality holds that facts are indisputable and can be divorced from values, and hence that scientific evidence has no wider symbolic meaning. Political rationality, on the other hand, deals with what is good and bad for people generally and for certain client groups specifically. It thus has both a moral and an interest-serving component within which "facts" may be interpreted in a variety of ways. In this context the following quote by a scientist is significant.

> "The adversary method for arriving at truth on which our legal procedures are based is, in simple language, not appropriate for arriving at sound public policy on scientific matters. Scientific matters simply cannot be settled by persuasive argument. The only effective method for resolving safely questions in nuclear or biological research is the objective analysis of experimental results by our best scientific minds .. The use of the adversary legal process to control scientific research is likely to lead to serious scientific errors and badly thought out policy" (McGill, 1978).

This distinction between the two forms of rationality is particularly troublesome when it is evident that scientists do not always agree on the nature of risk, and hence, on the appropriate regulatory standards that should be set, and even more disturbing when scientists enter the realm of "transcience" where experimentation and logical deduction are no longer scientifically possible. Transcience operates in those critical shadow areas between science and social judgement – the implications of low-level dosages of certain substances over large populations over long periods of time, the assumption that technological solutions can be found to control critically dangerous substances when no existing processes are commercially operable, the judgement that risks due to human failure or institutional incompetence can also be controlled by some technical means, and the whole evaluative process of comparative risk analysis and risk-benefit analysis alluded to earlier. All of these areas, so crucial in the current nuclear debate, cross the boundary between scientific evidence and political judgement. Yet it is not apparent that risk assessment techniques have yet adequately realized the genuinely difficult problems of reconciling the two methods of analysis and evaluation. The scientific and political controversy that followed the publication of the Rasmussen report on reactor safety (1975), and the Parker report on the Windscale thermal oxide reprocessing plant (1978), well illustrate these problems.

Risk Management and the Political Process

In many countries some kind of environmental impact analysis is now undertaken for major resource development (including energy) projects and risk assessment is growing in popularity. It behoves us to consider how these techniques will impinge on existing procedures for making decisions.

First of all, it is essential to relate these techniques to traditional "styles" of policy making. Broadly, there are two: the adversary approach most commonly found in the U.S. but also prevalent in Japan and West Germany and some other European countries, and the consensus approach followed in the U.K., other Commonwealth countries and in some European nations such as the Netherlands. (The countries cited do not necessarily follow one or the other approach entirely, but in practice tend to lean toward one or the other). The difference between the two styles is largely based on openness of debate and level of mutual suspicion. In the adversary approach the constitution encourages conflict and controversy which are often resolved by formal legal or political means, whereas in the consensus approach, policy is made incrementally through an elaborate process of consultation and compromise between the principal interested parties.

It is not surprising that EIA and RA have advanced further in those countries with an adversary style of policymaking which is based on challenge, conflict and the balancing of evidence, employing the courts to act as arbiters. The adversary "style" is dependent on widespread public information which is available to anyone. On the other hand, the consensus approach is based on co-operation among the dominant interests where little information is made public, and even less evidence is provided as to how decisions are reached, and where the courts play a muted role in soliciting evidence and resolving disputes. In countries following this latter culture of policymaking, EIA and RA are unfriendly devices for they endanger comfortable working relationships, open up avenues for interested groups to gain access to decision-making procedures and expose specialist advisers to contradictory information that could undermine their credibility. So it is doubtful that either EIA or RA will ever be welcome to those intimately involved in consensus-orientated policymaking.

Yet the paradoxical outcome of all these new techniques is that conflict is so likely and costly, and that time-consuming stalemates are so predictable, that in adversary-orientated political cultures, serious attempts are now being made to establish consensus-seeking mechanisms to facilitate the speedier and more felicitous resolution of disputes. For example, this is now evident in the efforts to establish a co-ordinated coal development policy in the U.S.A. (Carter, 1978), and it is apparent in attempts to streamline the nuclear plant licensing procedures both in Japan and the U.S.A. (and, one suspects, in West Germany and France). These efforts are taking the form of a series of task forces comprising the major interested parties (including representatives from environmental groups and consumer organizations) which meet over a long period of time and try to thrash out the principal points at issue. To date, few of these collaborative task forces contain politicians, which is a great pity, though in many countries, the legislatures have established various political scrutiny committees to review particular aspects of environmental policy, including energy policy. The usefulness of these committees, many of which are inadequately staffed and composed of people insufficiently informed and overworked is, however, of questionable significance in energy policymaking. At least this is certainly the case with the Commons Select Committee on Science and Technology in the U.K.

It is possible that over time, the merits of the two styles of political decision-making will be fused into a common approach. At present, however, there is still some distance to go, and in consensus-orientated political cultures the strains placed upon traditional decisionmaking procedures are beginning to show.

A major difficulty is that of ensuring the accountability of officials responsible for establishing risk standards, monitoring risk causes and enforcing codes of practice, namely those involved in the risk control function outlined in Figure 2. In the past it was assumed that these people could be trusted, for they were regarded as professionals proud of their responsibilities. Accordingly, the laws were framed to enable them to operate whilst protected by a cloak of confidentiality, and many officials were genuinely offended by even a hint that their work should be exposed to detailed investigation. In many instances, even politicians were unable to examine the assumptions upon which officials made their judgements and advised their respective ministers. But today, the tremendous publicity given to the failure or potential failure of technology and the upsurge of interest in RA has led to the suggestion that the custodians of public safety may not be able to do their job properly so long as they are protected by the laws of confidence. As a recent report written by the Council for Science and Society (1977) put it:

> "...we as a society are less well aware that impotence also corrupts, especially where it is linked to responsibility. An inspectorate that cannot enforce its requirements must either confess its impotence or conceal it from view. Thus a weak inspectorate is pushed towards identification with those who create the risks to the detriment of those that experience them."

It is difficult to suggest satisfactory ways in which accountability can be improved. The time-honoured British solution is to upgrade the accountability, not of the adviser, but of the responsible minister, through his answerability to Parliament. But this is now recognized as a hopelessly inadequate solution, as few ministers are ever compromised in parliamentary debate, and in any case it is the job of his civil servants to protect him from hostile questions.

Ideally, the basis upon which key decisions are made in the evaluative and control aspects of risk management should be made fully public and should be subject both to extra-parliamentary and parliamentary debate. This would allow different kinds of interest groups to have their say before final political judgements are made. There are various devices by which this could be achieved including a standing extra-parliamentary commission backed by an adequate research secretariat, and more powerful parliamentary investigatory committees. Properly speaking, civil servants should be publicly answerable to both investigatory bodies. Although, initially, this suggestion might lead to a lot of wrangling, it is quite possible that the American experience of working parties and task forces devolving from such organizations, but meeting from time to time in public and incorporating representatives from various interest groups, would quickly establish itself. This pattern of evaluative discussions might take a bit of time and cost some money, but by inducing a genuine sense of co-operation among all interested parties, it might well be more expedient, less expensive and arouse less suspicion than with the present methods.

Whether the two kinds of inquiry commissions could ever be merged – so that politicians sit on lay investigatory bodies – is a controversial question. Much depends on the political style of policymaking alluded to earlier. For example in the fairly open system in Sweden, an Energy Commission composed of politicians, trade union representatives, academics and informed lay citizens did debate various aspects of energy policy for almost a year. Though its recommendations were not binding in Parliament its findings obviously had a fair degree of political significance. In the more closed policymaking practice of the UK, on the other hand, the Energy Commission is largely an ineffective body noted more for the posturing of its participants than its constructive debate. Its direct influence on government policy is probably very small, but the fact that it has the ear of the Secretary of State for Energy cannot be ignored.

In general, it seems that even though we may be entering the era of the specialist politician who has to become expertly knowledgeable in only a few areas, the likelihood of politicians willingly sharing their decision making authority with lay interests is pretty small. Some kind of intermediate compromise along the lines already referred to may occur in exceptional cases (and here the energy field is an obvious candidate), but it is unlikely that either ministers or their senior civil servants will delegate much of their policymaking functions in the foreseeable future.

A second arena in which procedural adjustments will have to be made in order to incorporate EIA and RA into energy policymaking is that of adjudicating the degree of environmental protection necessary to permit specific projects to be constructed. Throughout the western world (at least in those countries where a degree of democracy exists) major energy projects are stalled or subject to cost overruns due to demands to reduce risks and minimize environmental impact.   The traditional device for resolving these disputes is the public inquiry, but as the Windscale example in the UK showed, the public inquiry is not suitable for this task for it is not structured in a manner to solicit evidence and balance risks against benefits.   Yet it is popular among bureaucrats and ministers for its rules suit those in power by controlling the degree to which the lay public have access to decisions.   Nevertheless in the light of developments in EIA and RA, the structure of the traditional public inquiry is unsuitable, whatever minor amendments are made, and so if it is not radically reconstructed, there will be more trouble.

No ready solution lends itself, but it has been suggested (Breach, 1978; Pearce et al., 1979), that there should be a pre-inquiry conference at which major points of dispute are aired and which can recommend a precise programme of research and investigation aimed at minimizing adverse consequences.   This could take the form of a task force or possibly an investigative tribunal of the kind proposed (but never tried) in the idea of the planning inquiry commission established under section 47 of the U.K. Town and Country Planning Act (1971). This would be followed by the public hearing proper at which only points of objection to the specific proposal would be raised (which could involve points raised from EIA and RA) and which could be incorporated in the conditions attached to planning consent.   Again, this would take time, but probably no longer eventually than would the unsatisfactory variations of existing procedures apparently so admired by senior government officials.

Should these twin suggestions relating to improvements in regulatory standard setting and enforcement of codes of practice, and to project appraisal, be converted into practice, then it is possible that the courts would become far less prominent as arbiters of risk acceptability.   This would be most advantageous since the courts should not enter the arena where complicated political and technical judgements have to be made, for lawyers and judges are neither scientists nor resource managers nor politicians.   The proper role of the courts surely is to ensure that legally defined rights are upheld and legally mandated procedures are followed.   The establishment of the former is a political responsibility and of the latter an administrative one.   The courts do have a crucial role to play where there is mischief and incompetence, but hopefully the suggestions for a more consensual, open approach to risk management should confine their activities to appropriate legal matters.

Conclusions

The major thesis of this paper is that the two analytical tools of environmental impact analysis and risk assessment, have encouraged spectacular and beneficial advances in the realms of identifying and estimating adverse consequences of resource development projects (and to a lesser extent resource investment programmes). This, in turn, has produced great anxiety among ordinary people uncertain of the implications of these findings, and consequently has placed some strain on customary procedures for defining acceptable standards, enforcing suitable codes of practice and determining the correct balance between environmental risk and social benefit. In short, both the evaluative and control phases of risk management have lagged behind the technically sophisticated identification and estimation phases. This in turn has led temporarily to an excessive dependence on scientific advice and a willingness to accept the particular rationality which underlies that advice. To combat these tensions, this paper recommends that both EIA and RA should be more carefully incorporated into policy review procedures as well as the more normal project appraisal process, so that the merits of certain courses of action and various alternatives can be more carefully considered before specific developments are proposed. The paper also recommends that new consultative mechanisms be established to ensure that those who are responsible for the risk control function are fully accountable, and that the environmental protection associated with particular major projects is properly analyzed and incorporated in the final project design. The key to these recommendations is the establishment of policymaking devices that merge scientific evidence with political judgement and enable interested and informed parties to participate fully in the advice upon which the peoples' representatives must eventually make decisions.

References

Anderson, F.R. (1973). "NEPA and the Courts: A Legal Analysis of the National Environmental Policy Act". John Hopkins Press, Baltimore.

Atherton, C.C. (1977). Legal requirements for environmental impact reporting. In "Handbook for Environmental Planning" (J. McEvoy and T. Deitz, eds), 9-64, Wiley Interscience, New York.

Breach, I. (1978). "Windscale Fallout". Penguin Books, Harmondsworth, Middlesex.

Burton, I., Kates, R.W. and White, G.F. (1978). "The Environment as Hazard". Oxford University Press, New York.

Burton, I. and Whyte,. A.V. (eds) (1980). "Environmental Risk Management" SCOPE 14. John Wiley, Chichester, Sussex (in press).

Caldwell, L.K. (1977). The environmental impact statement: a tool misused. (Unpublished paper) Department of Government, University of Indiana, Bloomington, Indiana.

Carter, L. J. (1978). Invoking the rule of reason - in the energy-environment conflict. *Science* **198,** 276-280.

Council for Science and Society (1977). "The Acceptability of Risk" Barry Rose, London.

Dreyfus, D.A. and Ingram, H.A. (1976). The National Environmental Policy Act: a view of intent and practice. *Natural Resources Journal* **16,** 243-262.

Fairfax, S. (1978). A disaster in the environmental movement. *Science* **199,** 743-748.

Kates, R.W. (ed.) (1977). "Managing Technological Hazard: Research Needs and Opportunities". Institute of Behavioral Science, University of Colorado, Boulder, Colorado.

Kates, R.W. (1978). "Risk Assessment of Environmental Hazard" SCOPE 8. John Wiley, Chichester, Sussex.

McEvoy, J. and Deitz, T. (1977). "Handbook for Environmental Planning". Wiley Interscience, New York.

McGill, W.J. (1978). Extracts from a speech. *Science* **198,** 275.

Otway, H.J., Maurer, D. and Thomas, K. (1978). Nuclear power and public acceptance. *Futures* **10,** 109-118.

Parker, J. (1978). "Report on the Windscale Inquiry". HMSO London.

Pearce, D.W., Edwards, L., Beuret, G., (1979). "Decision-Making for Energy Futures: A Case Study of the Windscale Enquiry". Macmillan Press, London.

Rasmussen, N. (ed.). (1975). "The Reactor Safety Study". Report WASH 1400, U.S. Nuclear Regulatory Commission, Washington, D.C.

THE PROBLEM AS SEEN FROM THE POINT OF VIEW
OF THE DECISION-MAKER

Carl Tham

Minister for Co-ordination and Energy,
Prime Minister's Office,
Stockholm, Sweden.

I have been asked to talk about the value of risk estimates seen from the position of the decision-maker. This is indeed a tricky theme and I must admit that I am perhaps even more confused about it after having read all the interesting papers for this Seminar. What can we do with all this impressive theoretical thinking about risk estimation? My reflections may be seen as a modest effort to evaluate the benefits and limitations of risk estimation from the perspective of a practising politician.

Let me start with a rather trivial statement: estimates of the risks and adverse health effects of different energy production systems are of course a necessary instrument in all decisions concerning energy policy.

It is also quite clear that the health effects and environmental impacts of energy production are now considered to be much more important than before, and therefore the importance of risk estimates has also increased. But let me add one thing. The intensive debate, particularly in Scandinavia, about the health impacts of energy production may very well have given many people a false impression that energy production is the main villain in our environment. The long recital of the risks imposed by different energy sources could reinforce that impression. But there are indeed many other – and perhaps more important – problems in the environmental and health fields – problems not connected with energy production As a decision-maker you also have to give a thought to the obvious benefits of a cheap and reliable supply of energy; and the risks also have to be compared along with other risks in society.

Cost-benefit analysis, in which costs also include estimates of the risks, could be one useful tool in the decision-making process. You could argue – and many of you certainly do – that all these estimates are more or less qualified guesses, that the uncertainty is quite considerable, but that is the way of most practical political decisions, after all; they *have* to be made, even when the facts are uncertain. The problem is very much the same, in say, the field of economic policy: the indicators, are unreliable, the future unknown and the advice is contradictory. Indeed, I would say that experts involved in estimates in the energy sector are generally more in agreement than those involved in economic forecasting. I think I have some practical knowledge of both categories.

The main problem with the uncertainty, seen from my point of view, is perhaps that we get more information on some risks than on others. Our knowledge of radiobiology and the nature of carcinogenesis is good enough to make quantitative estimates of carcinogenesis with some confidence even for low levels of radiation. On the other hand, the dose-response relations of chemical carcinogens are much less known; yet it may be that the dangers of cancer from fossil-fuels are greater than those from nuclear power. That is just one example; there are many others. The result may be that the decision-maker and the public pay more attention to the well-known risks than to those that are less well known, even though the latter may sometimes be far larger. Resources allocated to combatting these relatively small but recognised risks may thus be disproportionate when seen from a wider perspective – and the other problems are not sufficiently observed or discussed. A decision-maker has to be aware of this problem – and one conclusion is of course that scientific work in the more uncertain areas should be encouraged.

You can also argue – as indeed do several of the contributors – that all objective measurement of risk must involve some element of subjectivity and consequently must be biassed in one way or another. This type of problem recurs in many intellectual or scientific undertakings and, again, from the decision-maker's position, it is nothing new. He or she has to evaluate the quality of the information, including an assessment of the intellectual and scientific standard or reputation of the investigator or expert. I strongly dislike the current tendency to dismiss all information or knowledge by reference to the expert's background or employer. If there is a miscalculation or bias, deliberate or not, it should be shown by other estimates, and there should be no "guilt by association".

The widespread reaction against too much trust in experts and technicians, which we have experienced during the seventies, has in some circles developed into an extreme distrust of scientific and technical knowledge in general. Let us now use that dialectical process to establish a more mature balance between the two extremes – a balance which is very important in the field of energy policy.

Many times the problem is not the bias or subjective values of the expert but more that decision-makers and opinion-makers *use* provided information in a very selective way and with certain aims. It may be necessary to reach a decision, but it is dangerous if all the small print and reservations in risk estimates are forgotten or hidden. When reports of risks are misused the sinners are more often in the world of media and politics than in the realm of serious intellectual and scientific work.

My conclusion from all this is that the decision-makers could benefit from risk estimates in the energy sector. It would be a good thing if, at least in principle, all benefits and all risks connected with different energy systems could be put in one common system and decisions could be based on that. But there the real problems start. How do you measure the benefits of low energy prices and their impacts on industrial production and employment against, say, health risks from fossil fuel? The theoretical problems with this undertaking are big enough, which these papers clearly illustrate. But, worse, the political value of this kind of total assessment, in which you, indeed, combine hardware with software, is quite limited. Many of the more general problems and risks connected with energy production – its effect on the organization of the society, its benefits to many aspects of our welfare, etc. – are profoundly political problems, on which values can differ a great deal. The decision-maker is indeed better served with information which concerns the relative risks of *different energy sources providing the same level of benefits,* expressed in the same unit. Such rather traditional estimates are much more useful than speculative systems with a lot of parameters and implicit values.

It is sometimes said – and I have perhaps said it myself – that it would be desirable to develop a consistent risk philosophy as a tool for making decisions in a society. We have many risks, in all walks of life, and we obviously tend to estimate the importance of these risks in different ways in different contexts. Therefore, so the thinking goes, it would be fine to have a universal and reliable risk philosophy which could give us a basis for decisions in all fields. We could then operate in a more rational way when we try to combat risks to life or health.

Well, this is the assertion – but I must confess that I am increasingly doubtful of whether this is possible or, indeed, desirable.

It is of course true that we accept different levels of risks in different sectors. Risk estimates could also do a lot of good in making us more aware of this fact. But as soon as we try to find a common rational ground for all risk assessment, we disappear into a marsh of implicit values and theoretical and practical problems. It is necessary, for instance, to translate risks into costs for life-saving efforts and in that way put a price on life. This is now done implicitly, but I am not sure that we have so much to gain by explicit figures and a debate, among parties, organizations, experts and in the media, on the monetary value of life. Again, it might be a good thing to call attention to large risks for certain groups – occupational hazards and so on. But that is more in the field of risk identification, and not the same as creating a risk philosophy which is supposed to be a basis for rational decisions.

The futility of risk philosophies could be illustrated by the efforts within the late Swedish Energy Commission to establish a firmly agreed-upon risk philosophy in the field of energy. We had many distinguished and clever experts in the field at our disposal and they certainly did a good job. We could also agree upon some rather obvious things: that the energy system should not be allowed to give unacceptable consequences, that we have to avoid irreversible effects, that we have to think about coming generations and so forth and so on. It goes without saying that all these statements were vague indeed; and, as soon as we moved from the theoretical to the practical, we found that the meaning of all this good stuff was highly controversial. Some people used the "philosophy" to reinforce their disbelief in nuclear power, while others, a majority, drew the opposite conclusion, just to take one example. It could be said that this Commission and its members were especially ill-equipped for the task; but honestly speaking, I must say that the philosophies in other energy reports as well – at least the ones I have seen – are neither very illuminating nor suitable for practical purposes.

When you discuss the value of risk estimates you inescapably in the end splash down into politics: different values, ideologies and competition for power. And that means also, that you – as a politician – have to take into account not only what you think are the real risks but also the problems of subjective risks, the fact that many people seem to evaluate the risks in a way quite different from formal or objective risk analysis. That, if nothing else, has been made completely clear in the nuclear controversy. There is a gap between what most scientists believe to be the real risk and the general public's perception of the risk. The message is quite clear: to expect people's attitudes to a certain technology to be determined only by more traditional and comparative risk estimates is a super-rationality which could be irrational, at least for a politician.

In a democratic society no decision-maker can afford totally to neglect human feelings about risks and more subtle socio-psychological reactions. That is also an argument against an oversimplified use of risk estimates. Having said that, however, I would also like to stress that this fact does not absolve well-informed decision-makers or opinion-makers from the responsibility of informing the public about what they think to be real risks. Just to accept or, worse, even to foment, subjective risk perceptions without any effort to provide information about real risks is not acceptable.

So again, I am back to my first and rather trivial conclusion: politicians and decision-makers are best served with hard facts about risks and their impacts on health and environment. It then becomes our job to evaluate these facts as sensibly as possible and to make the necessary estimates of costs and benefits. I am not saying this to discourage either theoretical or scientific undertaking in the field of more sophisticated analysis, or efforts to develop more comprehensive risk philosophies. They could indeed be useful as background material or perhaps even as ideal models for rational decision-making. But we should not have any illusions that such efforts can easily be integrated into the work of the decision-maker.

THE RISKS OF ENERGY STRATEGIES: AN ECOLOGICAL PERSPECTIVE

John Harte

Lawrence Berkeley Laboratory, University of California, Berkeley, California, U.S.A.

Energy strategy assessment poses unique problems for the ecologist concerned with preserving healthy habitats and wildlife populations, and ecology provides a unique perspective for the energy-policy analyst. This paper will explore these two statements, provide examples to enhance their clarity, and discuss their implications for improvement of risk assessment and energy policy making. While the emphasis here will be on what is unique at the interface between ecology and energy-risk analysis, we in no way wish to suggest that what is not unique is not important. If ignored, what is non-unique may well lead to serious degradation of ecosystems; but it is emphasis on the novel that will help to illuminate the subject of risks and add new dimensions to the study of energy policy and perhaps even of ecosystems.

Some of the ideas developed here are adapted from an 800-page report to the U.S. National Academy of Sciences, entitled Energy and the Fate of Ecosystems (Harte, 1978), which was prepared by a group of 19 ecologists and energy specialists. It describes the conclusions of the Ecosystems Resource Group of the National Academy's Committee on Nuclear and Alternative Energy Systems (CONAES). The two-year study undertaken by the Ecosystems Resource Group looked at practically all possible technologies that could contribute to energy supply in the U.S.A. through to the year 2010, and assessed their ecological risks under a wide variety of assumptions about their deployment. In addition, the current and potential ability of ecology to provide the kind of predictive capability needed to set rational energy policy was assessed. Some discussion of the difficulties involved is provided by Harte and Jassby (1978) and by D. Goodman (1976). Conversations with many colleagues have benefitted my thinking on the subject of risk assessment. John Holdren and Robert Socolow, especially, will recognize their own contributions herein.

The Energy-Ecosystems Interface: Three Propositions

Study of the ecological risks of alternative energy strategies
has suggested certain conclusions that may have bearing on how
risks are perceived, evaluated, and incorporated into the making
of energy policy.   These conclusions are discussed below under
three propositions:

Proposition 1

Long causal pathways and numerous feedback loops characterize
the energy-ecosystems-human welfare triad;   ecosystems amplify
and stretch out in both space and time the risks and impacts
of technological activities.

Ecological impacts of energy activities are rarely confined
to the immediate place and time of the activity. The consumption
of water from the upstream reaches of river networks can exert
profound influences on estuaries, and on people who derive benefits
from those estuaries, possibly hundreds of kilometres downstream.
Chronic deposition of toxic metals in lake sediments and watershed
soils can create a "time-bomb", causing impacts decades later
if a pH reduction in runoff waters makes these metals soluble
again.   We have found Figure 1 to be a useful way to structure
terminology and thought around the complexities of environmental
risk assessment. As shown in the figure, abstractly, and as illustrated
by example in Figure 2, the risks of energy activities do not terminate
with ecosystem damage.   Human life and health is also linked
to the existence of healthily functioning natural ecosystems
(Westman, 1977; Harte, 1978; Harte et al., 1978a; Holdren 1978).
Table 1 lists some of the amenities provided by healthy ecosystems

TABLE 1

Goods and Services Provided By Ecosystems

1.	Capture of solar energy
2.	Storage and cycling of nutrients
3.	Water storage and control of water flow
4.	Climate and weather control
5.	Purification of water and air
6.	Conversion of photosynthate to higher trophic levels
7.	Pest control
8.	Provision of wilderness for seclusion, beauty, recreation
9.	"Refining" chemicals (natural pesticides, rubber, drugs, etc.)
10.	Maintenance of a genetic "library"

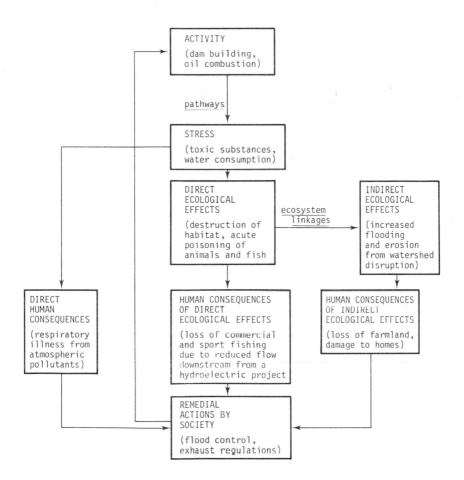

Fig. 1 The stages of ecological impact. Energy-related activities (such as dam building, fossil fuel burning, or slurry pipelining) put stresses on ecosystems. These stresses can be direct (e.g. toxic action on organisms). Because of the complex interconnections among ecosystems, they may also lead to longer-lasting indirect effects. The stresses can also directly affect humans. Because natural ecosystems provide people with many goods and services, the indirect ecological effects can also adversely affect society. Response to the loss of goods and services is often some remedial action which initiates another pass through the sequence.

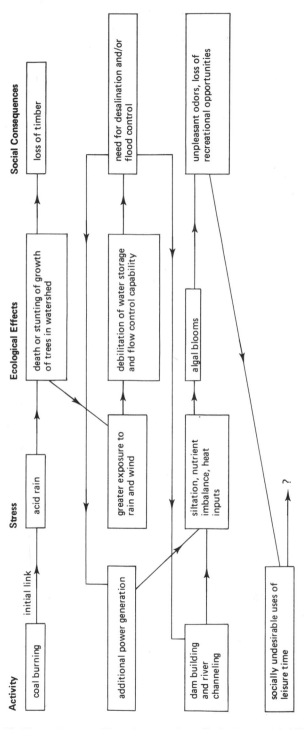

Fig. 2 Environmental impacts of some human activities. Some of the potential interconnections among activities, stresses, ecological effects, and social consequences are illustrated for the particular case of an initial stress consisting of acid rain. Even for this simple example, the interconnections shown are both incomplete and speculative.

TABLE 2

Consequences of Loss of Ecological Amenities

Society forced to rely more heavily on:

1. Industrial fertilizers, pesticides and herbicides for agriculture

2. Irrigation water for agriculture

3. Wilderness lands for agriculture

4. Desalination or other treatment for water

5. Erosion and flood control measures

6. Artificial climate control (air conditioning, humidifying, etc.)

7. "Plastic" for entertainment

Society forced to contend with:

8. Greater fluctuations and/or decrease in supply of foods and fibre

9. Odours and other aesthetic affronts

10. Increased incidence of diseases and outbreaks of noxious pests.

11. More unpleasant weather

12. Poorer air quality (e.g. due to impairment of vegetative filtering)

13. Poorer water quality or reduced quantity

14. Diminished natural diversity

15. Fewer options for future innovation and improvement of life

16. Increased ecosystem instability

17. Greater vulnerability to fire or flood

---

and Table 2 lists some of the consequences to society of the loss of ecological amenities. Two problems for risk assessment can now be posed: understanding the sensitivity of these naturally-provided amenities to human activities, and evaluating the loss to humans when these amenities are degraded.

Traditional environmental impact assessments have largely ignored both of these problems, focussing instead on the link between activity and stress, and on the direct effects of stress on ecosystems (Harte, 1978; Harte and Jassby, 1978; Holdren, 1978). What little ecological literature exists on the subject of sensitivity of environmental amenities to ecosystem disruption is rarely up to the highest standards, being largely a product of speculation and of consulting firms hired to prepare environmental impact reports.   Progress is necessarily slow because advances here must build upon firm ecological insights.

Increased efforts in fundamental ecological research, directed at the generic system level (e.g. the estuary or the watershed) are particularly needed (Blanchard, 1978; Harte *et al.*, 1978a). Examination of the historical record of ecosystem destruction and its human consequences could augment such efforts, providing a vast wealth of case-study information.   In contrast are studies oriented around energy activities, which attempt to analyse all of the risks associated with a particular activity such as strip mining or dam building. While useful for certain purposes, this approach often becomes bogged down in the complexities of the early stages of risk assessment (viz the links among activity-stress-direct effects) and rarely confronts the problems of environmental amenities. The multiplication of uncertainties along these first stages of risk analysis often destroys the public's confidence in the ability of experts to predict precisely what ecological degradation will take place.   The result is that further analysis of amenity disruption is dismissed as overly speculative.

Evaluating the loss to humans which arises from environmental amenity disruption is an equally challenging facet of risk assessment. Economic analysis is useful occasionally in providing lower bounds on the dollar costs of such disruption but may also retard progress towards enhancing sensitivity to intrinsically non-monetary and multi-faceted benefits of healthy ecosystems. Most needed now, in the absence of adequate ecological and institutional insights into the problems of understanding and evaluation, is an awareness on the part of decision-makers that the risks to humans of ecological degradation are incompletely understood and almost certainly underestimated.   It may, indeed, be the case that no rational process (i.e. one in which the rules are agreed upon in advance) can ever substantially reduce the evaluation problem, both because of the diversity and often intangible nature of ecological benefits to man, and because of difficulty of finding agreement about what constitutes welfare. Under *Proposition 3*, below, we return to the problem of evaluation and offer some suggestions.

Figure 1 suggests another problem for risk analysts.   Human society can respond in a variety of ways to the loss of goods and services that results from ecosystem degradation.   These responses often unleash new stresses and therefore new risks which exceed the risks of the original loss.   Thus a complete risk analysis of the effect of strip mining in an arid region cannot end with a description of what the land will look like after reclamation is attempted, nor can it even be content to describe the hardships of reduced water availability for downstream users.   It must include an assessment of how society is likely to respond to those hardships and assess the risk of *those* responses. This is an enormous job, rarely even attempted in impact assessments.   Whether the feedback loops even converge rapidly enough to allow rational analysis is uncertain. Lack of insight into the nature of the goods and services provided to man by natural systems makes progress difficult here, but parallel research could clarify how society responds to losses of naturally provided goods and services. Economists, along with sociologists and perhaps even psychologists, could exert a major influence here in the study of risk. Not to undertake this responsibility is to remain looking only at the tip-of-the-iceberg of environmental and human risks.

The *magnitude* of the impacts of energy activities on ecosystems, rather than anything qualitatively new, brings this issue to the fore in the context of energy-policy risk analysis. It is particularly important when dealing with *ecosystem degradation,* as compared, say, with *direct human health* impacts of energy, because of the frequently observed necessity for high technology (dams, pesticides, energy for climate control) to restore some measure of the welfare removed by disruption of environmental amenities.

Proposition 2

The concern of ecologists mainly with habitats and entire wildlife populations rather than with individual organisms introduces novel uncertainties in risk analysis.

In the study of human-health risks of energy, it is possible to associate pollutants with diseases on a statistical basis (either epidemiologically or with the use of laboratory animals) without having even the remotest causal understanding of the mechanisms involved. Because concern over human health is specifically a concern with human mortality and morbidity rates, the information provided by these studies can in principle assume a form useful for risk evaluation. The unit of concern is the individual human, and these units come in numbers large enough (per class of environmental circumstances) to allow progress to be made. In ecology, the unit of concern is whole habitats and wildlife populations rather than individual organisms. These units of concern are usually either unique, or so rare, as to preclude controlled studies. Or they are too valuable to manipulate in the name of research. Or they are so complicated that the proper analogues of morbidity and mortality (which, of course, even in epidemiology provide a woefully incomplete picture of human damage) cannot be measured or even selected. Or they are not even spatially, temporally, and taxonomically defined.

The sensitivities of whole populations to the demise of some fraction of its individuals are generally poorly known, and thus dose-response information for individual organisms is not always very useful in an ecological context. Here and there great progress has been made using statistical correlations and/or controlled manipulations. A statistical study (Vollenweider and Dillon,

1974) of the effects of nutrient loading on the potential for lake
eutrophication demonstrated that the individuality of lakes does
not preclude the possibility of deducing systematic response
to external disturbance.   Another classic study (Likens *et al.*,
1978) of watershed manipulation demonstrated the particular
effects of the clear felling of timber on nutrient storage capability,
and more generally demonstrated the potential for controlled
manipulations of natural systems.   But generally the goal of either
a statistical or predictive understanding of the response of ecosystems
to anthropogenic stress remains elusive.

The differences between the unit of concern in ecology and
in epidemiology creates an additional problem in risk analysis,
as illustrated by attempts to assess the risk of habitat degradation.
Consider, first, the consequences of polluting the air above city
A and then in addition polluting the air above City B, where A
and B are, say, 200 kilometres apart. To a good first approximation,
the health effects are additive.   If 100 additional deaths per
year can be expected in A and 150 additional deaths per year
are expected in B, then the combined risk is well-summarized
by stating that 250 additional deaths per year are expected.
This is not at all the case with ecosystem degradation.   Many
forms of habitat, such as wild rivers and virgin forest, are sufficiently
scarce and delicate that highly non-additive effects can be expected
if these systems are subject to environmental stresses.   The demise
of wildlife species and the greatly increased travel time imposed
upon people seeking recreation in these habitats are just two
examples.

Compounding these problems are the uncertainties inherent
in complex systems.   By inherent we mean that further research
cannot possibly reduce them.   As was shown first in meteorology
(Lorenz, 1969) and then later extended to ecology (Platt *et
al.*, 1977), there are limits to the ability to predict the behaviour
of complex systems. These limits arise because of the extraordinary
sensitivity of such systems to initial conditions.   Small errors
in the initial specification of, say, the state of the atmosphere
or a marine ecosystem can grow so fast in time that detailed
prediction beyond a few weeks would require an initial-condition
data base so large as to exhaust the capacity of even the
largest computers.   With replicable units as in epidemiology,
statistical information can substitute for predictive understanding
based on models of causal mechanisms.

For the energy-policy maker the implications of the above
should be sobering.   Some uncertainties are here to stay and
decisions will have to be made in that light.   Thus, the notion
that we can "build it now and figure out what it will do to us
later" is often wrongheaded.   Moreover, the inability to properly
conceptualize the units at risk leads to the inability to portray
to the public the type of risk created by an energy policy decision,

let alone the probability of its occurring. Thus so many of the ecological impacts that have attracted attention in the past few decades came as nearly complete surprises to all. For example, the near eradication of certain top carnivores as a consequence of DDT use could not have been anticipated by scientists viewing the system-at-risk from the pesticide to be farmlands and peripheral ecological communities. Is it not likely that further surprises await us in the future? Perhaps the steady chronic leakage of petroleum into freshwater lakes and streams, a process going on wherever automobiles are in use, will be seen someday to have greatly altered aquatic life.

For the ecologist, the implications are equally profound. Single organism dose-response studies have provided considerable information about the sensitivity of organisms to pollutants. But in addition to the problems mentioned above, this approach suffers from the disadvantage that it cannot include the synergistic and compensatory effects known to exist in natural communities or populations. A particular species of fish may not be directly affected by a given level of a pollutant, but it may, for example, undergo in the wild a population explosion if the food supply of its competitor is.

Awareness of the inadequacies of single-organism studies and the difficulty of controlled field manipulations has lately led to interest in laboratory ecological microcosms which can be manipulated cheaply (for example, by the addition of an energy-related pollutant) and controlled. While causal relations are far easier to deduce under these conditions, than in the field, it is not yet clear that the relations so deduced are relevant in a natural context (Jassby et al., 1977, Harte et al, 1978b). More effort is needed to fully understand and exploit this potential for risk assessment.

Proposition 3

The large variety of types and consequences of ecosystem degradation, (Proposition 1) and the high degree of uniqueness of ecosystems (Proposition 2) combine to exacerbate a problem present in every facet of technology assessment — the problem of non-commensurate alternatives.

Trade-offs frequently occur when some facet of human welfare is improved. For example, the provision of more energy entails increased risk to human health and increased degradation of environmental quality. Perhaps less obvious is the observation that attempts to reduce one kind of environmental risk often increase environmental risks of another kind. For example, one approach to reducing thermal pollution in streams and lakes is to require that electric power plants be cooled by wet cooling towers rather than once-through methods. Such a requirement is, in fact, built into the U.S.A. Federal Water Pollution Control Act of 1972. However, this reduction in thermal pollution entails a large increase in water consumption by evaporation from the wet cooling tower. The consequent reduction of stream flow may lead to downstream effects on aquatic life which exceed those caused by thermal pollution. Quantitative comparison of trade-offs of this sort can be difficult to perform and can frustrate efforts to determine whether we are better or worse off as a result of what was intended as an ameliorative measure (Budnitz and Holdren, 1976). Part of the difficulty lies in the fact that we often do not know enough about the sensitivity of ecosystems to particular stresses to allow an accurate characterization of the effects. A part of the problem is that, having characterized effects on each side of the trade-off, we do not know how to reduce them to common values or units so that we can compare them. (Goodman, 1976; Harte and Jassby, 1978).

This problem which we call the comparability problem, is not unique to ecological impact assessment. A piece of technology for improved automobile fuel combustion can lead, as history has shown, to other detrimental health effects. The comparability problem is perhaps more manageable here, however, because common units – mortality and morbidity rates – are available to allow comparison. Admittedly, determining the trade-offs in these rates may be a difficult task, but a fairly well-defined measure of concern to the public does exist. In contrast, ecosystem degradation can assume a bewildering variety of forms. And its consequences for human welfare are also diverse and not reduceable to a few simple dimensions. How do we compare the nearly certain extinction of a flower or an insect with the less probable extinction of a higher mammal? More practically, how do we compare the ecological effects of strip mining and deep mining, or of wet-tower cooling and once-through cooling? Neither dollars nor deaths and disease conceivably can provide a valid unit of comparison for ecological trade-offs and it is doubtful that any such unit could be found. The prospects for an ecological oracle which would perform input-output calculations to sum up effluents and other stresses, deduce the ecological impacts, and then pick the ecologically most appropriate energy strategy from a choice of strategies seem dim, indeed.

In the absence of an ideal "scientific" solution to the problem of comparability, improvements can still be made by reducing certain institutional barriers. Institutional improvements are needed both to facilitate understanding and communication of risk, and to facilitate decision-making directly. We discuss the former here and the latter in the Conclusion.

One greatly needed improvement is to learn how to speak to one another about ecological risks with greater forthrightness. This case has been made carefully and eloquently elsewhere in the context of a debate over a hydropower and flood control project on the Delaware River (Socolow, 1976). Certain practices all too prevalent in environmental risk analysis impede progress here. Frequently one sees statements of the type: "The water consumed annually by a proposed coal-slurry pipeline to transport U.S. Western coal will be only one-hundredth that used annually for irrigation by farmers in the Great Plains." Or in the same spirit, the probability of death from a release of nuclear materials from the uranium fuel cycle is compared with the probability of death for Californians from an earthquake. The deception infused into risk assessment from such comparisons results from attempting to compare the risks of activities which have non-commensurable benefits. (Budnitz and Holdren, 1976).

Risk assessment focussed on comparisons of comparable risk of various pathways of energy strategies for achieving common benefits is likely to be most illuminating. An example from the CONAES report illustrates this approach (Harte, 1978; Harte and El-Gasseir, 1978). There are many ways to use coal. It can be burned directly for some uses, or it can be converted to electricity first, or it can be converted to gaseous or liquid synthetic fuels which can either be used directly or converted to electricity. Because different pathways for using coal may be optimal for different end uses or benefits, overall assessment of the optimum pathway is likely to be a futile effort. The CONAES report considered a particular end use – that of home heating. Ruling out direct coal burning in the home for obvious reasons, electric heating (with the electricity produced either from coal directly or from synthetic coal-derived fuels) versus direct heating with synthetic fuels are the choices. Which of these is most appropriate? Here again, the problem is too broadly stated to allow a simple answer. The consequences of the pathways will vary with the type of ecosystem and will very likely be non-comparable. A more manageable comparison can be achieved by looking at a particular type of risk. The CONAES Ecosystems Report focussed upon a comparison of water consumption in each of the coal pathways for home heating. While water consumption is not a fully understood measure of impact, it does encapsulate a large number of effects. Water consumed is water unavailable for further human use. Water consumed from a flowing stream means an alteration of aquatic habitat, probably a raising of water temperature, and, therefore, probably a lowering of dissolved oxygen levels. As long as the same benefits are being compared (in this case the heating of homes) comparison of this common measure of risk is useful. Because water consumption at widely disparate sites can have non-comparable consequences, such comparisons are ideally made on a regional basis taking into consideration regional differences in water supply and its fluctuations (Harte and El-Gasseir, 1978).

Water consumption for home heating is of course only one example of a "common-benefit, common-risk" comparison. We submit that the total effect on overall energy strategy risk analysis of having a large number of such comparisons for various risks and various benefits would be far more useful than what would emerge from the same amount of effort put into optimization programmes based on cost-benefit analyses and attempts at inter-risk and inter-benefit analyses.

In order to make it easier for the public to assimilate the information that will emerge from the type of risk assessment advocated here, it is important that the variables used be standardized as much as possible and that they be appropriate to the complexity of the problem being described.

The case for standardization is straightforward. Standard units or measures such as kilometres per litre (as an index of performance in vehicular fuel consumption), or coliform counts, (as a measure of health risks from water), or cubic metres per gigajoule (as a measure of water consumption caused by energy activities), or gallons of heating oil per degree-day (as an index of performance in buildings-energy conservation), provide a common language with which people can talk to one another about environmental stresses, impacts, and consequences.

A standardized, or commonly accepted, variable which is inadequate or misleading can do more harm than good, though. If the variables do not reflect properly the complexity of the facet of the world they are meant to describe, then progress towards thinking about risks is surely impeded and wrong notions become dogma. Many health specialists think, for example, that coliform counts are a misleading index of the health menace of water for swimming or drinking. Similarly, energy conservationists are not agreed that the degree-day is the most appropriate indicator of absolute fuel need for space heating or cooling. Enough is understood today about the chemistry and effects of air pollution to know that it is deceptive to express particulate emissions in a unit that lumps together particles of all sizes.

There is today a considerable lack of appropriate indices describing fluctuations of environmental parameters. The stresses resulting from energy activities rarely act in isolation from other natural stresses. Thus, the ecological effects of water requirements for cooling power plants or reclaiming land after strip mining may be insignificant in years of normal rainfall and devastating during periods of drought. Damage to terrestrial vegetation from acid precipitation may be particularly severe when the metereological conditions conducive to very low pH precipitation coincide with periods when vegetation is particularly sensitive such as germination, flowering or budding (Likens, 1978). The traditional notion of "threshold", a notion easily converted into legislative standards, is inappropriate in circumstances such as these, for it is implicitly a one-dimensional measure. The real threshold for unallowable water consumption may be quite high in wet years, and quite low in dry years. Every Nth year on the average, (where N is very large) the threshold might be zero. If an industry is to consume a fixed quantity of water year after year, how can one set a threshold?

A start along these lines requires appropriate indices of natural fluctuation (Harte and El-Gasseir, 1978). For the case of stream flow, we have advocated the use of a two-variable index, denoted xQy. This quantitiy is the x-day, y-year low flow, and is defined as the lowest flow rate, averaged over x consecutive days of the year, expected on the average every y consecutive years. The quantity xQy is important both to the water consumer and to the ecologist studying impacts of the consumption. For the water consumer it allows calculation of the amount of water storage needed to survive periods of drought, or it can be thought of as a measure of the degree of flexibility in operating schedule needed to accommodate to droughts (for example, it tells how often and for how long it is necessary either to switch to more expensive, but less water-consuming, modes of cooling or to shut down operations altogether and transfer customers to other suppliers). For the ecologist concerned with the maintenance of riparian and estuarine habitat, the double degree of freedom in xQx is also important. Simpler indices, such as the z-per cent flow (i.e., the flow exceeded all be ($100-z$)per cent of the time) do not easily make contact with information about the tolerances of aquatic organisms to periods of uninterrupted stress. Because xQy provides information about both the duration of the drought (x) and its frequency (y), it is a richer statistical measure of flow and one upon which site-specific water consumption criteria for industry can be constructed.

Another area in which standardization of appropriate variables will facilitate risk analysis is in the growing use of microcosms for ecological impact assessment. It is hoped that eventually microcosms will provide a convenient, manipulable, and replicable means of making impact assessments; as yet they are in their infancy and methods for their use have not fully developed. Because they are artificially assembled systems, it is possible that impact studies using microcosms will take on a chaotic pattern, or it is possible that standardization will occur.

The identification and adoption of standards for design, initiation, and operation procedures for microcosm use will be of enormous benefit, as it will facilitate communication among microcosm researchers and enhance the possibility that experimental results can be duplicated. Many of the advantages of microcosm studies over field research will be lost if such standardization does not take place.

Moreover, it is apparent that care should also be taken that the variables selected to be measured in the monitoring of microcosms are optimum and that the form in which data is presented is standardized and readily communicable. Given the usual constraints

under which experimental science is performed and given the
need for relatively frequent monitoring of microcosm properties
in ecological research or environmental impact assessment, it
is important that measurements be inexpensive enough and quick
enough to be performed regularly on an array of systems. Moreover,
the variables to be measured must be chosen so that the problem
being studied is elucidated.

While no single list of parameters can suffice for all purposes,
even if attention is limited to just one type of laboratory ecosystem
such as freshwater lakes, it might be very helpful to determine
and agree upon a minimal list of parameters – those which are
necessary for drawing and duplicating conclusions based on microcosm
studies.

Conclusion

Finally, we must confront the most difficult question of all.
How should society make decisions about energy policy – decisions
that will affect the fate of ecosystems and therefore human
welfare – in the light of all the difficulties described above?
Even assuming there takes place improvements of the sort advocated
here, in the style and substance of ecological research, risk evaluation,
and communication with the public, people still have to make
policy decisions. As we have stressed, these will inevitably be
made in the face of uncertainty. Are there ways in which the
decision-making process, itself, can be adapted to the constraints
imposed by ecological realities?

One way to begin thinking about decision-making is to divide
it into two stages: option defining and option selecting. A common
childhood crisis arises when a piece of pie has to be shared by
two children. An ingenious and exquisitely fair solution to this
problem in decision-making consists of having one child divide
the pie (i.e. set the options) and the other select the first piece.
Imagine how cumbersome it would be to do it the adult way,
with a trial cut made in the piece of pie, followed by a cost-benefit
analysis of a particular assignment of the pieces. Interestingly,
a simple algorithm exists for generalizing the child's pie-sharing
game to an arbitrary number of sharers. Can that approach then
be used to improve resource allocation and energy-policy-making
procedures? Obviously it cannot if taken too literally. In a battle
over whether or not to dam a river, conservation tactics such
as dividing the river laterally down the middle would leave the
dam builders no effective choice! But the inherent fairness
behind the principle of separating the selection of options from
their definition suggests that it somehow ought to be reflected
in energy-strategy decisions.

A problem with the formulation of energy policy today is
that these two processes, defining options and selecting options,

often are not separated and instead are handled by the same interest group. Examples are: the past choice of the type of nuclear power reactors in the U.S.A., the choice of shipping routes for transport of oil on the high seas, and, at least in the past, the choice of design and location for major hydroelectric projects. In some cases, a small measure of constraint on option selection is exercised by the public in the form of regulations, as illustrated by the recent U.S. surface-mining legislation which restricts future coal surface-mining sites according to criteria based on slope, precipitation, and present use of land. A second example is wild-rivers protection legislation which forbids development on specified rivers. The ecological benefits of introducing even a small degree of separation between definition and selection of options are unquestionable. Can the degree of separation be enhanced?

At the risk of appearing extremely naive, we imagine structuring the process of making decisions about energy in the following stages:

1) General goals might be set by the political process (an example is the goal of providing 1000 mw of additional electric power in a certain region in a certain time period) and all feasible ways to reach the goal (including conservation efforts to liberate wasted electricity) would be described in detail.

2) At this stage, major public interest groups would reduce the options listed under 1). above according to various environmental, public health, and socio-economic criteria.

3) The reduced list would be given to the energy industry or utility for final selection based on criteria of their choosing.

Clearly, if such a scheme were tried, many unexpected problems would arise. The question of just who gets to limit options in step 2). is sure to pose conflicts. The approach might in fact be a lot worse than the less structured, less participatory, way we proceed today. Perhaps constructive collaboration of environmentalists and industry in both definition and selection of options, as in the very innovative National Coal Policy Project (Murray, 1978), is a better approach. Greater public discussion of reform in decision-making could help greatly to clarify these and other choices. If this discussion does not take place, the principle of separation will be violated by default, even in the very process of selecting decision-making methods.

References

Blanchard, C. (1978). "The Costs of Environmental Degradation" (master's thesis). Energy and Resources Group, University of California, Berkeley.

Budnitz, R. and Holdren, J. (1976). Social and environmental costs of energy systems. In "Annual Reviews of Energy",

(J. Hollander, ed.), Vol. 1, 553-580. Annual Reviews Inc., Palo Alto, California.

Goodman, D. (1976). Ecological expertise. *In* "Boundaries of Analysis" (H.A. Veiveson, F.W. Sinden and R. Socolow, eds), 317-360. Ballinger, Cambridge, Massachusetts.

Harte, J. (ed.). (1978). "Energy and the Fate of Ecosystems". Report of the Ecosystem Impacts Resource Group of the Risk/Impact Panel of the Committee on Nuclear and Alternative Energy Systems, National Academy of Sciences/National Research Council, Washington, D.C.

Harte, J. and El-Gasseir, M. (1978). Energy and water. *Science* **199**, 623-634.

Harte, J. and Jassby, A. (1978). Energy technology and natural environments: the search for compatibility. In "Annual Reviews of Energy", (J. Hollander, ed.), Vol. 3, 101-146. Annual Reviews Inc., Palo Alto, California.

Harte, J., Blanchard, C., El-Gasseir, M. and Tanenbaum, S. (1978a). "Environmental Consequences of Energy Technology: Bringing the Loss of Environmental Services into the Balance Sheet – Part 2: Services, Disruptions, Consequences". Draft Report to the U.S. Council on Environmental Quality, Energy and Resources Program, University of California, Berkeley, California.

Harte, J., Levy, D., Rees, J. and Saegebarth, E. (1978b). Making microcosms an effective assessment tool. *In* "Proceedings of the Symposium on Microcosms in Ecological Research", Augusta, Georgia. (in press).

Holdren, J. (1978). "Environmental Consequences of Energy Technology: Bringing the Loss of Environmental Services into the Balance Sheet – Part 1: A Framework for Analysis". Draft Report to the U.S. Council on Environmental Quality, Energy and Resources Program, University of California, Berkeley, California.

Jassby, A., Dudzik, M., Rees, J., Lapan, E., Levy, D. and Harte, J (1977). "Production Cycles in Aquatic Microcosms". EPA Office of Research and Development Report No. EPA-600/7-77-097, National Technical Information Service, Springfield, Virginia.

Likens, G.E. (1978). Acid precipitation. *Chemical and Engineering News* November 22, pp29-44.

Likens, G.E., Bormann, F.H., Pierce, R.S. and Reiners, W.A. (1978). Recovery of a deforested ecosystem. *Science* **199**, 492-496.

Lorenz, E. (1969). The predictability of a flow which possesses many scales of motion. *Tellus* **21**,(3), 289-307.

Murray, F.X. (ed.). (1978). "Where We Agree: Report of the National Coal Policy Project". Westview Press, Boulder, Colorado.

Platt, T., Denman, K.L. and Jassby, A.D. (1977). Modelling the productivity of phytoplankton. *In* "The Sea: Ideas and Observations on Progress in the Study of the Seas", (E. Goldberg, ed.), Vol 6. Wiley, New York.

Socolow, R. (1976).   Failures of discourse.   *In* "Boundaries of Analysis: an Inquiry into the Tocks Island Dam Controversy" (Feiveson, H., Sinden, F. and Socolow, R., eds.), 9-40. Ballinger, Cambridge, Massachusetts.

Vollenweider, R.A. and Dillon, P.J. (1974). "The Application of the Phosphorus Loading Concept to Eutrophication Research". NRC Associate Committee on Scientific Criteria for Environmental Quality. National Research Council, Ottawa, Canada.

Westman, W.E. (1977).   How much are nature's services worth? *Science,* **197,** 960-964

POLICY ISSUES IN STANDARD-SETTING: A CASE STUDY OF NORTH SEA OFFSHORE OIL POLLUTION

David W. Fischer

*Institute for Industrial Economics,
Bergen, Norway.*

The standard-setting process has both a socio-economic and a political basis. This has been stated previously by Majone and Holden. Majone (1975a, 1975b) noted the following points in line with this thesis

- a standard cannot be set on a purely scientific basis;
- a standard provides only an appearance of precision and hence of "scientific" character;
- a standard always represents an implicit evaluation of human well-being;
- a standard is only one of several alternative means of regulation;
- the institutional framework often determines the decision on a standard;
- self-interest of regulatees moves them to attempt to modify the terms of the regulator, including any standards set.

In addition, Holden (1966) noted that regulatory processes, including standard-setting, are based on a bargaining process between regulators and regulatees. As an example of the political nature of standard-setting Schon (1971) notes that an attempt to set standards in the lumber industry for the "2x4" developed more political response than any other issue in the recent history of the U.S. Department of Commerce. Lumber producers both large and small, building interests, federal agencies, state governments as well as U.S. Congressmen and Senators were all attempting to influence the thickness standard of the "2x4". Schon posits that a system of "dynamic conservatism" builds up around an existing technology to protect it whenever it is threatened with, say, standards that might initiate any loss of market, influence or position.

A characteristic of energy production as an issue which arouses and focuses attention on standard-setting is its location at a particular site.   Both the need for energy development and its attendant benefits and costs are perceived in local terms, even though such development has large overall national benefits and costs.   What heightens the energy issue even more at the local level is the use of nuclear power, which is not restricted by location but is shiftable among a variety of locations, while petroleum or coal may often be developed *in situ* where it is extracted. Wherever a nuclear power station is placed both benefits and risks are heaped on a particular site, regardless of the overall benefit-cost ratio to the nation.

Both the site and the energy technology appeal to the "protective instincts" of the particular social groups or "actors" that array themselves around each.   Those local actors who stand to gain from the development of energy at a site ally themselves with those interested in embedding and extending that energy system in the overall economy.   Those local actors who do not perceive direct economic benefits or do not hold values compatible with that energy system (say, nuclear) will see themselves as bearing the total risk of such a development and oppose it.   Those orientated to the site tend to be motivated by traditional values well embedded in the existing social system.   The premise for conflict tends to be that if those in the locality do not pursue their own interests, then their interests will be ignored by those at the national level, who are more orientated to the technology than to the site. What precipitates conflict is a specific proposal to place a nationally approved technology on a particular site. This places site-orientated and technology-orientated values into a conflict where each set of actors attempts to protect the values inherent to the social system surrounding either the technology or the site. This attention to a specific project at a specific site tends to shortcut broader discussions of goals and alternatives.

Technologies, Environments and Threats

Figure 1 is an attempt to display the basic processes involved in the protection of a technology and of a site.   It does not try to model the protection process *per se*;   but rather attempts to simply portray an overview of responding to a threat to some technology through an effort to regulate it.

At any one time society has both energy and environmental values and needs that must be met through whatever technologies and sites that exist. The existing energy production system embodies all of the technology and social value systems necessary to meet the energy demands of the economy it serves. The existing environmental system also embodies all of the technology and social value systems necessary to maintain the local employment and living base over time at the existing locale.   In the case of environmental values such awareness can grow around a traditional technology supporting

ENERGY SYSTEM   **ENVIRONMENTAL SYSTEM**

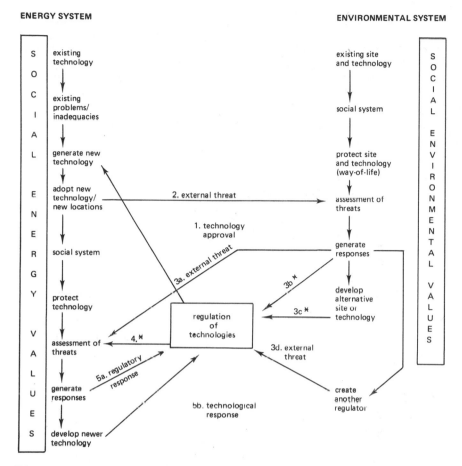

Fig. 1 Threats and responses from technologies associated with different social values.   3b* = regulatory response;   3c* = technological response;   4* = regulatory threat.

some existing economic activity on a site sustaining a way of life such as single-family fishing. The existing technology becomes traditional through the social system that builds up around it at that site.   Often a community will be entirely based on it. Because this technology is closely tied to the society it supports, that society becomes protective and develops some capability for determining threats to that technology and hence to their way of life and overall environmental values. Should any external threats be perceived or felt, responses are generated in either a direct threat to the offending parties, a direct approach to the regulator concerned, an alternative technology for meeting the problem (which, in turn, must be approved by that regulator), creation of a new regulator or some combination of all of these responses.

Existing technology can be creating problems or developing inadequacies in fulfilling society's energy needs and values. To meet the perceived energy demands or resolution of these problems a new technology is generated through research and development of some pilot plant. Over time this technology is then approved by some regulator and becomes the new way of meeting society's energy needs. At this point a new social system develops in the energy bureaucracy, as well as in supporting institutions, to maintain the status of this adopted technology. The protection of this technological system is fostered through an assessment capability that determines whether threats, either external or internal, exist and generates a set of responses. If the threat is seen as external, from another social-technological system, based on, say, environmental values, then responses are formed to counter this threat. The regulatory body that approves and authorizes technologies of, say, energy is often the focal point for such conflict. Responses can be external to the regulator – a direct attempt to fix the existing regulatory stance or provide a new technological capability for meeting regulatory shifts. Threats can also be generated within the energy system from some competing energy source.

The kinds of responses that affect the energy system and its regulator from those with, say, environmental values, include some of the following:

- change society's energy values (external threat);
- ban the new technology (external and regulatory threat);
- pursue adoption of an alternative technology (external technological threat);
- attempt to create another or change the regulator (external threat to regulatory and energy system);
- create new standards for technology (technological threat);
- change existing standards for approval and acceptance (technological threat).

These responses are often attempted in concert, and it is interesting to note the role of standards. Either setting new standards or changing existing standards can provide major channels for those opposed to energy values already accepted by those in the energy system as well as those regulating it.

Those in the energy system also respond to such threats; however, these threats are normally confined to the regulator, with whom they have evolved a rapport. Such responses include:

- counter environmental arguments directly to regulator;
- ask for special standards to be created;
- ask for a tightening of existing standards;
- generate newer technology internally.

Again standards can play a role in the responses of those

supporting a, say, large energy system.   Again the response is either to set standards where none existed or to tighten existing standards as a means to forestall further actions from environmentalists and others with allied values.

The Concept of Actors

From Figure 1 and the above discussion it is readily seen that a system of protection does develop around a set technology. This protective system includes a wide range of actors that have essentially the same set of values or core beliefs in the technology as a means to achieving their economic and personal goals.   Such actors for an energy system would normally include:

- industry characterizing the energy source;
- supporting industry, including contractors and suppliers;
- supporting unions, workers and company towns;
- supporting research and development institutes;
- supporting journalists and publications;
- government regulator;
- allied government units, including energy, finance, and industry;
- allied international governments and agencies;
- affiliated financiers;
- infrastructure suppliers and contractors;
- public utilities;
- consumer associations, including industry, commerce and home;
- political representatives associated with the energy source;
- local governments in towns centred around the energy source.

As can be readily seen, this actor array includes a wide spectrum of support for an energy system.   These actors while apparently unlinked, would unite around an energy technology that meets their prevailing energy and economic values and needs.   Some external threat to the energy system will bring together these seemingly disparate parts into a cohesive system for protecting the technology on which all of them depend.

The actors as listed above indicate the potential behavioural structure between the energy system and some other system. For example, environment as shown in Figure 1.   This actor linkage and interplay can be manifested in a variety of ways (Fischer and Keith, 1977).

The interface between the energy system and some other system, say, environment, is complex, including numerous actors and decision-makers with diverse, indirect and subtle positions, and often conflicting objectives.   Therefore, when considering individual measures such as environmental standards many of these actors can be regrouped to demonstrate only the interactions and interests at hand.

Figure 1 ends at attempting to display an energy technology

threat via some regulatory process. Figure 2 is an attempt to
show the actor array involved in the standard-setting process.
This figure links the groups expected to participate in regulation
of a technological system.

Figure 2 can be seen as basic to the case reported here. An
energy system defined on this basis shows the minimal actor

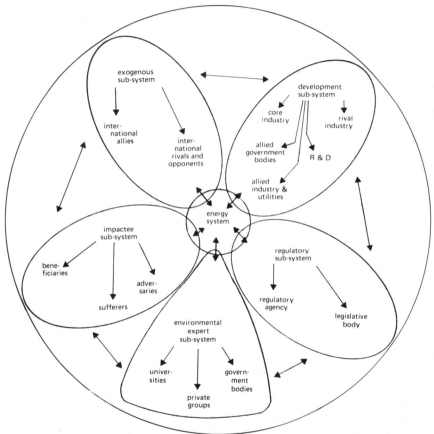

Fig 2. The actor sub-groups involved in an energy system

configuration required for an adequate solution to some problem
of interest. Each of these actor sub-systems comes into the
energy system with different values and for different reasons,
such as:

Development actors:

value – profit or service potential;
reason in system – generation of energy via approved technology;

Regulatory actors:

value – reconcile opposing demands and approve technology;
reason in system – legislative/administrative authority;

Environmental expert actors:

value – professional and personal interest;
reason in system – recognition of energy impacts;

Impactees:

value – preservation of environment/livelihood;
reason in system – dissatisfaction with potential outcome;

Exogenous actors:

value – preservation of environment/energy systems;
reason in system – treaty, international influence.

Each actor sub-system can be seen as a set of individual units that goes through some decision-making process in its particular decision field. Placed together these units provide a systems perspective. A rational approach can be embedded in each sub-system but it does not follow that a completely rational solution will be sought in the overall system, much less that there will be agreement between sub-systems based on different technologies and social values.

Each sub-system provides different foci, incentives, information, and alternatives for the different clusters of actors within it. These inconsistencies are often concealed, ignored or set aside whenever a threat emerges to that actor grouping. At such a point the choices made are usually normative rather than analytically based. Thus while a rational analytic perspective is useful it does not provide a holistic framework for describing the interplay among actors.

Standard Setting and Actors

In a complex and large-scale energy programme such as the North Sea petroleum effort, few actors were found with unique or single-dimensioned values. Indeed, many central actors have attempted to become so broad in outlook as to make the decision-making system appear even more diffuse. The defining of the boundaries of the decision problem and the means for addressing separate policy issues, such as in standard-setting, thus emerge as central issues.

One clear observation is the deepening of values over time as the particular energy programme proceeds. Determining a stable development path for both the economy and the environment such a complex array of value positions can prove difficult. The time horizons of such values cannot be wholly relied upon for guidance of policy responses, as both short and long-term contradictory objectives and expectations may exist within the same value set – such as short-term local public revenues from oil companies versus long-term economic dependence on them.

Interplay of actors, and therefore interplay of controls such as standards, is often indirect. For example, ecology itself is not of direct interest to corporations but environmental goals influencing company capital and operating costs are. These in turn are of no direct importance to environmental bodies. Translating actors' values into environmental standards cannot be done directly without the intermediate steps of identifying impacts associated with the various stages of the proposed development programme. Next, these impacts must be arrayed against the actors involved to determine their degree of interest and influence in changing the outcomes. Finally, the mechanisms for change or acceptance must be seen between all such actors in both policy systems.

Actors tend to become involved in a specific site development issue if they can expect to have some significant impact on the outcome. Those actors with local ties to the site tend to focus on those avenues giving weight to the local point of view, such as their legislative representatives, while actors having direct ties with the technology emphasize institutions weighted toward technology, such as the regulator or a specialised legislative committee or agency.

Establishing the focus of the energy system is in itself a strategic consideration. Technology-associated actors are often served by a narrow problem or issue definition which reduces the amount of information necessary, deflects attention from broader issues, precludes participation of certain actors and obscures connextions with alternative means. Environment-orientated actors tend to define the issues more broadly, to include social and economic goals in conflict with those served by the offending technology, and otherwise to open up the decision-making process. Purely site-orientated actors focus on their economic livelihood and traditional lifestyle and tend to use the traditional political process to influence the regulator.

Policy Issues

Some of the issues one could expect in a North Sea Oil pollution study would include those noted in Table 1. The table suggests key elements in the standard-setting process. Information is grouped under key actors and choices, associated choices and hence associated actors, choice criteria, uncertainties and conflicts.

Regulators, developers and environmental experts have roles more important than those of the impactees, while exogenous

TABLE 1

Summary of Offshore Oil Discharge Standard-Setting Process

	REGULATORS	DEVELOPERS	IMPACTEES	EXPERTS	EXOGENOUS
Key actors	UK: PPD, DOEn Nor.: SPCA, Min. of Env.	– oil comps. – poll'n control eq. mfgrs.	– fishing industry	– poll'n control research – fisheries research – marine research	– EEC – others
Key areas of choice	– set standards – meet production goals	– implementation of standards – eq. alternatives to meet standard	– effects of standards on catch – change fishing pattern	– standards alternatives – how monitor effect of standard	– set constraints on national standards
Associated choices	– condition of development (exemption) – monitoring eq. – poll'n control eq.	– purchase of eq. – monitoring eq. – R & D	– monitor catch	– costs of standards – tests of eq. – monitor ambient env. & samples of fish	– effects of standards on trade & development – quality of North Sea env.
Associated actors	– Energy – Industry – Finance	– other oil comps. – other eq. mfgrs.	– other fishing nations	– food, fisheries – other researchers	– non-EEC countries
Choice criteria	– acceptability of standard to comps. – government image & obligations	– cost of implementing standard – cost of maintaining relationships – R & D costs	– sales volume – way of life	– public reactions – eq. capabilities – research capabilities & resources	– EEC agreements – other nations agreements
Uncertainties	– standard's value to env. – standards impact on comp. – poll'n ctl. eq. availability	– basis of standards – future government standards	– future market value & volume	– claims of mfgrs of poll'n ctl. eq. – effects on env. – non-technological means for poll'n control	– effects on trade – effects on env.
Conflicts	– internal: production vs. env. – external: informal, closed vs. formal, open decision process – requests vs. requirements	– cost vs. need – standards vs. guidelines – maintain relations vs. fight standards	– access to decision process	– researched advice vs. best guess – gaining research funding support – env. perspective vs. technology perspective	– international vs. national emphasis

eq.mfgrs.= equipment manufacturing/manufacturers

actors provide general constraints to the national systems. The impactees who have the greatest need for standards and are part of the reason for the existence of such standards have the smallest role in setting them.

"Muddling through" still characterizes some of the actions of industry and government, such as industry providing no alternatives to a proposal or government merely reacting to industry initiatives, both of which only reinforce traditional patterns of assessment and response. A common pattern is to drift into joint solutions to environmental problems between industry and government where circumstance and expediency require or justify action, even before the inherent environmental complexities are made clear. The wide array of actors and potential actors to become involved in energy-environment responses appear to play a lesser role than would seem warranted judging from the lack of environmental information and alternatives known to exist. Drifting into joint solutions among core actors with wide discretion and informal relationships can distort the broader process of decision-making where representative bodies should play a major role. Because the government (majority party) can define the terms of its relationships with outside interests, actors and pressure groups, there is little requirement for these outside-actor groups to be concerned with or go through Parliament. Such decision strategies vis-a-vis other actors raise issues that affect the regulatory or standard-setting system itself. The results of a systems approach would not show in the immediate future where core actors dominate the decision-making system and attempt to influence other parts of the system to their viewpoint. While it is clear that other actors are becoming more vocal, the actual persuasiveness of these other actors is still open to question.

Policy questions suggested by the case study of standard-setting for oily water discharges in the North Sea include the following:

1) Who sets discharge standards?
2) Who is excluded from standard-setting?
3) What is the quality of the information base, and how is it obtained?
4) What is the actual basis for setting standards?
5) Are standards effective in reducing oil pollution?
6) What value is analysis in standard-setting?

These six questions form the basis of the remainder of this paper and are presented in the form of the actual process involved along with the issues generated by that process.

Energy Versus Environment – who sets the standards?

The governments of both Norway and the United Kingdom (U.K.) have attempted to organise their environmental standard-setting process to interconnect with the offshore developers. The main difference between the U.K. and Norway in the regulatory actors involved in standard-setting is that in Norway the central actor is the SPCA,[1] an environment actor; while in the U.K. the main responsibility is in the hands of the PPD,[2] an energy actor (aided in its research by the CUEP,[3] a unit in the Department of the Environment). Thus the U.K. has placed the standard-setting process for offshore environmental protection directly into the agency responsible for overseeing continued petroleum production from offshore. Norway, on the other hand, has sought a counter-balance – an environmental agency instead of an oil production agency regulating oil company pollution. Figures 3 and 4 show the interface between the energy and environment policy systems in the U.K. and Norway. The differences between the two patterns are readily evident. For a more detailed treatment of the process involved, see Fischer and von Winterfeldt (in 1978).

U.K. officials claim that the actual placement of the standard-setting agency is of little consequence on the standards eventually set and enforced because of their consensus style of government. Any standard set is the result of a massive in-house co-ordination effort. Norwegian officials note that the source of the expertise used is important in offsetting any possible bias. Indeed, formal co-ordination is required under the procedures used by Norwegian officials in passing on all petroleum-related offshore development. While no actor is capable of holding back or otherwise influencing the issuing of the drilling permit, no energy actor can influence the standard set for oily water discharges.

Whether the placement of the standard-setting and implementing agency is relevant or not may be resolved by looking at the standards actually set. The standards in the U.K. at present are 30-40 ppm as an average of two daily samples, to be exceeded only 4 per cent of the time up to a maximum of 100 ppm. This is a guideline standard used to set individual standards that can vary for each platform approved. Norwegian standards were set at around 50 ppm and later reduced to 25 ppm as an average based on continuous sampling, not to be exceeded more than one a month. No maximum discharge level was specified. Norway uses no specific guideline standard but treats each platform application on a case-by-case basis.

This sheds little or no light on the problem of who should set and enforce standards. The U.K. has more experience in setting standards as it has eight production platforms operating in its sector. With two exceptions for smaller platforms, the guideline standards have been applied. Norway has only one field producing at present. Its three platforms discharge oily water at around 100 ppm, exceeding the

---

[1] SPCA = State Pollution Control Authority
[2] PPD = Petroleum Production Division
[3] CUEP = Central Unit for Environmental Pollution

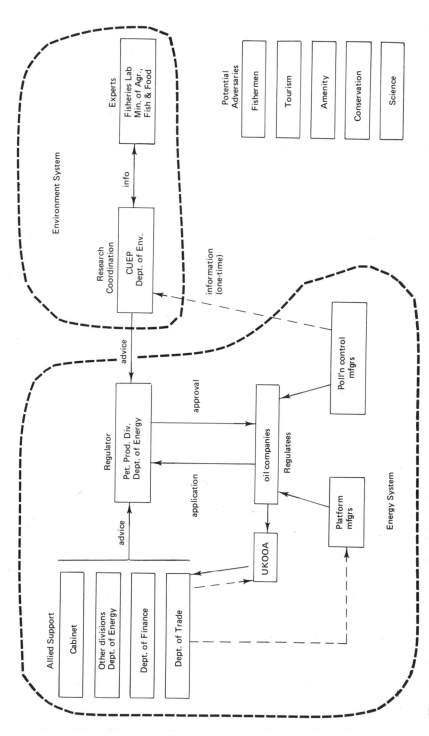

Fig. 3 An assessment of the energy-environment interface in the UK for setting standards for offshore chronic oil discharges. UKOOA = U.K. Offshore Operators Association.

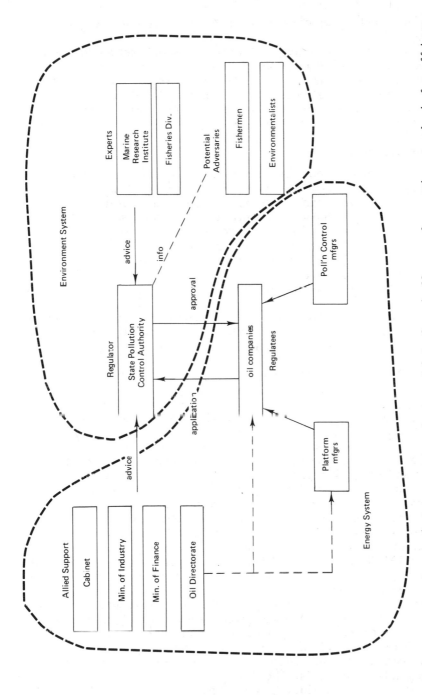

Fig. 4 An assessment of the energy-environment interface in Norway for setting standards for offshore chronic oil discharges.

existing standard which was set after these platforms commenced operation.

Criteria that could help determine who should set standards might include the quality of information available on expected impacts and damages as well as the potential for protracted conflict with impactees or those affected. As it turns out both of these criteria emerged as issues in the standard-setting processes of the U.K. and Norway.

Who is Excluded?

The oil companies and platform operators are in direct contact with the regulation agency (SPCA in Norway and PPD in the U.K.) on questions of oily water discharges from a specific production platform. The companies submit a design proposal or equipment specification with operating characteristics to the regulator for approval. This is then reviewed, and either accepted or modified. Furthermore, the SPCA and PPD can make contact with treatment equipment manufacturers when it comes to specific questions relating to treatment or monitoring equipment for a particular platform.

In the actual setting of the standards, regulatory authorities however made little contact with oil companies. In the U.K., regulatory officials made some informal contact with oil company representatives, but in each case the contact was to provide advance information on the forthcoming guideline standards rather than to seek company input for the standard-setting process. In addition, the CUEP sponsored a seminar to inform oil companies of the basis for their guideline standards. Company officials expressed concern over the apparent lack of interest in both sponsored marine research and views on the degree of severity of the standards set. In Norway, contact with oil companies is made during the standard-setting process because it is handled on a case-by-case basis. However, Norwegian officials stress that they set the standard and the oil company only furnishes background information.

Various governmental and non-governmental experts are involved in the standard-setting process in both Norway and the U.K. The Sea Fisheries Laboratory of the Ministry of Agriculture, Fisheries and Food (MAFF) provided the CUEP with expertise about the possible effects of chronic oil pollution on fish and other marine organisms. The Warren Springs Laboratory provided technical expertise on oily water treatment. However, university researchers and expert councils, such as the Nature Conservancy Council or the Natural Environment Research Council, were not directly involved in the analysis of chronic oil pollution by the CUEP.

Experts are to some degree involved in the case-by-case standard-setting in both countries. The PPD sends the discharge application of the operator to various governmental groups, including MAFF,

Warren Springs Laboratory, and the Marine Division of the Department of Trade, where they are reviewed and commented on. The SPCA sends applications for discharges not only to governmental experts, but also to fishery union experts and non-governmental environmental experts.

The main difference here between the U.K. and Norway is that the PPD relies solely on in-house expertise, while the SPCA also includes experts from outside of government. The confidentiality of the operator's report and the desire to maintain a good working relationship were mentioned in the U.K. as the main reasons for keeping the review within the governmental agencies.

Potential sufferers from chronic oil pollution include fishermen, consumers of fish, coastal residents, tourists, the tourism industry, and local governments. A different type of impact may be felt by ecologists, conservationists and environmental groups, whose values – protection of rare species, birds, etc. – are affected.

These impactees had a possibility to be heard at a public seminar held by the CUEP, but they were not directly involved in the process of elaborating discharge standards. In the review process by the PPD, the impactees were not involved, although some governmental experts could be interpreted as representing their personal or professional interests (e.g. biologists from the MAFF). In the Norwegian review process some impactees were involved; for example, fishery organizations receive discharge applications for comment.

Exogenous actors play a more limited role in the routine review of discharge applications. The PPD and the SPCA do not inform each other about every discharge application. There are contacts between the SPCA and the Nordic environmental convention, but it is not clear to what extent these contacts include discharge application reviews.

In both the U.K. and Norway, links in the energy-environment systems seem to be lacking in the private sector, both within the energy system and between the two systems. In the energy system the largest constraint to pollution control equipment installed on offshore platforms is lack of space. Pollution control is seen by oil companies and platform manufacturers as an "add-on" rather than an integral part of a platform design and operation. It would therefore seem important to establish a link between platform manufacturers and pollution control equipment manufacturers. This link should serve as a means of integrating designs and operational capabilities, thereby reducing constraints and costs of pollution control.

Another missing link seems to be that between the private pollution control equipment manufacturers and the environment system. A strong link here would provide better information about such equipment to those advising the regulator. In addition, environmental research findings could be given to the pollution control equipment manufacturers for redesigning performance capabilities. Subsidies for development of new approaches in the private sector could also be fostered.

Fishermen and environmentalists have less direct and, indeed, more tenuous links to the regulators. The decision to install a platform is made entirely within the energy system, with no environmental inputs as to location or design. Only the add-on technology of pollution control is the issue to the regulator, oil company and manufacturer.

Thus the spectrum of actors included in standard-setting does not adequately span the actor array as suggested in Figure 2. Figures 3 and 4 show the general linkages between the actors described above.

Significant sub-groups of actors, including oil company representatives, treatment equipment representatives, non-governmental experts and impactees are excluded from the process. The main difference between the U.K. and Norway appears to be in Norway's greater willingness to co-ordinate discharge applications with non-governmental actors even though the information gained may be of low significance. One reason for this more open process appears to be a hedge against possible complaints about unpredicted oil-pollution impacts.

Few actors can directly identify with the offshore environment. No actor group actually lives in the North Sea and few direct links of oil pollution damage can be traced to an offending platform. Therefore, no direct and continuous complaints from excluded actors can be heard in official circles. While both countries claimed that offshore oil pollution was not a serious problem, still they were the first countries to set strict discharge standards. The exclusion of certain actors from this process appears to be more the result of the style of decision-making than of deliberate intent.

Whether or not the actors excluded from the standard-setting process affected the standards finally set is moot. Significant information was probably not missed but standards were set without much notice given to affected parties. Some university researchers in both the U.K. and Norway were concerned that they had been given no opportunity for input. In Norway any actors having an interest and wanting to provide information can do so providing they know that an application is pending and how to go about giving this information. In the U.K. this opportunity does not exist. As no adequate and formal procedures exist for assessing the regulatory processes in either of these countries conflict could be expected should excluded actors experience oil pollution damages.

Environmental Research - the information base

The quality of biological information available for setting standards for oily water discharges from offshore platforms left much to be desired. Adequate marine research prior to the need to set environmental standards simply did not exist for the North Sea. In view of this, how can standards even be set? Oil companies noted that their experiences and independent research from the Gulf of Mexico indicated that discharges around 100 ppm or more did no long-term damage to the offshore environment. On the other hand, non-oil producing countries bordering the North Sea were taking a very conservative view of potential oil pollution from platforms. In addition, Norwegian fishery exports were of such economic significance that a demand for strict discharge standards existed. Finally, the U.K. was subjected to pressures through the EEC, which preferred rather strict across-the-board standards. Only adequately researched information on the North Sea environment could provide the key to the resolution of conflicting demands. Known oil pollution effects are summarised by Cowell (1977), who notes that the effects of oil pollution on the marine environment are less than expected.

While both countries have adopted monitoring programmes to support their standard-setting process it is clear that information on the long-term effects of chronic oil discharges on marine organisms is inadequate. One would like to answer such questions as: Given normal fishing operations and no oil development, what is the amount and quality of fish catch that the fishing industry can expect? To what degree do different levels of chronic oil pollution reduce this? To what degree does chronic oil pollution endanger the ecological balance in the marine environment?

The basic answer is: we do not know. There seems to be an enormous uncertainty about the biological effects of oil discharged at low but constant levels into sea-waters. Ambient oil concentrations are hard to detect against background hydrocarbons in the sea. Even at a highly polluting platform in the Ekofisk field (at least 10-30 tons of oil are being discharged there monthly) which is located near a very large herring spawning ground, marine biologists cannot claim that there are effects of oil on fish.

The process linking emission to the fish or marine organisms in the food chain is uncertain as the fate and turnover of oil once released is currently beyond measurement. But in the judgement of marine biologists, effluent with 50 ppm oil discharged offshore in reasonable amounts with good dispersion will not lead to any direct harmful effects on fish or other marine organisms. On the other hand, they do not reject the possibility that small amounts of oil pollution may affect behaviour (e.g. spawning) or have long-term cumulative effects on fish and in the food chain. An additional source of uncertainty arises onshore, where synergistic effects of oil together with other pollutants may occur. Norway in particular did not give major credence to biological research done in other countries. This finding is surprising because of the general lack of both information and agreement on interpretation of research results.

The high uncertainty about biological effects created another conflict among the actors involved in the regulation problem. Experts from the offshore operator's side claim no effects whatsoever with present treatment levels.   Some marine biologists and fishery representatives warn of potential long-term effects.

The main result of this uncertainty was in the Norwegian standard set on a rather pessimistic view.   The pessimism with respect to consequences has, of course, several other supporting sources: e.g. the strong Norwegian sentiments favouring fishing and the general policy of slowed development of the oil fields.   The first makes a pessimistic assessment of biological effects a political necessity; the second allows the pessimism to be translated into rather strict standards and regulations.   Practically, this pessimism is reflected in the policy of using the best technical (applicable) means to prevent chronic oil pollution.

In the U.K. regulation effort, the uncertainty about biological effects resulted in a shift from environmental considerations to equipment performance and cost considerations.   Rather than starting with a worst environmental case attitude U.K. regulators asked which treatment could be implemented without imposing too high a cost to the developer.   This consequence of the uncertainty about biological effects is compatible with the U.K. objective of rapid oil field development and the general environmental policy of using "best practicable means" to reduce pollution.

Monitoring is used in the U.K. to provide a data-base for current standard-setting.   Oil companies are given responsibility for such monitoring.   Each platform is to be monitored twice a day by the company having operational responsibility.   The PPD determines the methods of monitoring and analysis.   Twice a month the platforms are inspected by the PPD, but since helicopter space must be reserved in advance no surprise inspections are possible.   The results of such monitoring can be subjected to analysis by a government laboratory, at the expense of the company involved.

In Norway, oil companies are asked to do baseline monitoring at the platform site before construction occurs, in co-operation with the Marine Research Institute.   During production, monitoring continues at fixed points around the platform, the company again taking primary responsibility in co-operation with the Marine Research Institute.   Only levels of oil discharges are collected.   These figures and samples are then sent to be evaluated by the Marine Research Institute and the Fishery Research Institute.   The former institute also conducts an independent monitoring programme on oil levels monthly.   The complexity of hydrocarbon compounds and the relationship of oil with other pollutants makes evaluation of monitoring results (as well as setting standards) difficult.   Also, companies are reluctant to do continuous monitoring and then send such data to independent institutes for fear that their own figures would be used against them.

Thus, although the lack of environmental information may have had an effect, the standards derived were more strict than elsewhere. The important point, however, is their lack of a biological basis, demonstrated by the fact that the actual quality of the environment was not considered in setting the discharge standards. In no case discussed did environmental quality appear to be a significant factor. There appeared to be no fundamental change in the location, design, technology or operation of any offshore project based solely on environmental quality criteria. Rather in each case standards were tied directly to performance of available treatment equipment. Data other than that for treatment equipment covered physical aspects of the North Sea, such as winds, currents, etc., rather than biological or ecological conditions.

Considering this uncertainty and the importance of this issue, one would expect both countries to have started comprehensive marine research programmes on effects of oil pollution. Although the problem was known years ago, up to now the research response has been slow and *ad hoc*. Both oil company and university environmental experts criticize the respective government marine research efforts for inadequacy as well as lack of co-ordination.

Standards Set - what basis in practice?

Since they lack information on biological and chemical effects of oil on the marine environment, regulators in both countries have turned to practical solutions. Both countries rely wholly on entrepreneurial endeavour for oil discharge treatment alternatives. Neither country is attempting to develop treatment alternatives, and only the U.K. tests such equipment as available. The standards as actually set depend directly on the treatment equipment:

- availability: existence;
- capability: performance;
- practicability: size and price.

One important factor is the platform space available for such equipment, which in the extreme was reported to approach ten per cent of total platform cost. Given this high cost and the lack of data on oil in the marine environment it is surprising that the oil companies have been so willing to install treatment equipment. It would be very difficult for the regulators to make their case on a strong research basis.

Since Norway has opted for a case-by-case standard-setting process, oil companies are unable to predict acceptable treatment measures. Norway has favoured stronger standards by relying on *ad hoc* technological innovations available from treatment equipment entrepreneurs. At any point in time the operator may opt for either existing technology or waiting for another advance. No guidance is given for this choice although a penalty could accrue should a given technology be chosen too soon. The basic guidance available is from the manufacturers of such treatment technologies, who are prone to exaggerate the performance characteristics of their respective designs.   An operator can also decide between undertaking his own research in treatment technology or investing in a given techno-logical design from another source.   Once an operator has invested in a given design he has an incentive not to continue research which could render his own investment obsolete or influence a further tightening of standards.

Since standards are based solely on existing treatment equipment, little incentive exists for innovation or for expanding the information base.   The link between released oil and damages does not have to be specified because existing technology reduces the problem to some lesser local magnitude.   This has also made it possible to ignore such alternatives as recycling, reinjection or reprocessing.

Neither the U.K. nor Norway has gone beyond this reliance on existing technology, simply because there is little demand for them to do so.   The regulators have successfully imposed this technology on the oil companies despite a poor data base, and no strong voice exists for them to go further at this time.   The issue of oil pollution is a recurring one, however, due to frequent accidental oil spills.   While regulators and oil companies prefer not to see a link drawn between accidental and operational oil losses it is obvious that both sources contribute to a growing oil pollution in the North Sea.   Independent researchers and fishermen have been quick to point to this overall problem but they have not yet had much influence.

Effectiveness of Standards – reduction of pollution

There is little question that the standards set in both the U.K. and Norway will be effective in regulating oily water discharges to specified levels, provided that believable sanctions exist.   Surprise inspections are nearly impossible because of the need to reserve space for helicopters.   However inspections do occur, and the treatment equipment is expected to be operating.   Daily reports on discharges are expected by the regulator even though such monitoring is done solely by the oil companies.   Some provision has also been made for *ad hoc* ambient monitoring by government marine research groups, which allows for some independent checks.   Discharge standards do not seem to have operated as a constraint on oil production, so compliance would appear to be no great problem.

Given the enormous need for revenues by both the government and the oil companies as well as the lack of a well-researched link between released oil and damages, there does exist an incentive not to stress a strict adherence to such standards.   At present, however, there is no evidence that either country has not responded to this potential oil pollution in an acceptable manner.   In Norway with its important economic interest in fishery exports there is a strong political voice to ensure that oil pollution reduction is accorded high priority.   In the U.K., offshore economic interests are much smaller and thus the dependence on the goodwill of public servants is much higher.

Oil pollution in the vicinity of the Ekofisk Field in the Norwegian sector is reported to be higher than the adopted standard. However, this standard was set after this field went into operation and there is an expectation that eventually the platforms will install acceptable treatment equipment.

Neither the U.K. nor Norway ever considered other means, such as emission taxes or bans, to reduce chronic oil pollution. The problem was simply to bring a potential oil pollution source under some regulatory standard. A charging scheme would probably not be effective, since the revenues so greatly exceed the total costs of extracting oil. It would seem that oil companies pay any cost necessary to extract oil, if one looks at the enormous risks and expenditures incurred in offshore exploration. Bans on operations would appear ineffective given the strategic and economic importance of oil. It has been reported that the U.K. temporarily halted production at one of its fields for environmental reasons, but the priority given to oil can be seen in the fact that production at Ekofisk was underway again within four hours after the blowing well at Bravo was brought under control.

Analysis - Assessment Methods

Methodologically, where does one begin the assessment of standard-setting? What approaches appear viable to either those responsible for setting the standard or those studying it independently? Standard-setting is not a neat and simple problem, as our research has shown. No one decision-maker exists, but rather a plethora of actors each confronting the other with different values and opinions and fuelled by the substantial uncertainty on the effects of offshore oil pollution.

Noting the political, institutional and non-scientific nature of standard-setting, we investigated the problem from four perspectives: environmental economics, policy analysis, decision theory and game theory.[4] Obviously, a major constraint in our approach included the scientific personnel working on this project, and the above approaches reflect our training, experience and bias.

Environmental economics was quickly discarded as it soon became obvious that economists have not devoted much attention to standard-setting. Rather they use a charge scheme based upon expected theoretical behaviour. Pearce's summary of the standard versus charge problem concludes that either may be applicable depending on the case (Pearce, 1976). Suffice here to say that every problem noted for setting a standard can be equally applicable to

---

[4]See "Decision-Making in the Environmental Protection Agency", (Committee on Environmental Decision-Making, 1977). This volume contains an extensive study of U.S. environmental regulatory decision-making, including a survey of applicable methods for analysis.

setting a charge and every advantage claimed for a charge can be attributed to a standard (Majone, 1978). Cost-benefit analysis was also discarded because allocative efficiency was not the objective in either country, nor was it possible to determine willingness-to-pay estimates that would even approximate to usefulness (Fischer, 1975a and b). Since Norway and the U.K. had already opted for discharge standards our goal was to study this decision and its outcome as a process.

Policy analysis was used as the basic approach in determining just how standards were being set. This approach as described above was based on the author's earlier case studies involving multiple actors with conflicting values and inadequate information (Keith et al., 1976). The main advantage of this approach is the structuring of the problem into its perceived issues for subsequent analysis by a variety of methods and processes. For example, this analysis identified problem areas and provided basic inputs into the decision- and game-theory models. It also aided officials in viewing their tasks as a systems process and thereby provided them with a perspective for widening their efforts. Finally, the framework used suggested a basis for facilitating decision-making.

Decision theory also aided in structuring the problem. Decision trees were applied to actor objectives and alternatives. The actor policy framework of regulator-developer-impactee provided the basic components to a decision model to generate alternative standards based on the probable utilities of each actor group for each of their alternatives. Von Winterfeldt (1978) has made a major application of decision theory to the North Sea Oil pollution case. Given the structure of his model, the performance uncertainty of the treatment equipment and the regulator's inspection procedure were dominant features in defining the range of standards.

Game theory was used to attempt to account for the multistage process in standard-setting via multiple interactions among the three actor groupings. The model finally derived was not applied because of its simplifying assumptions. It can, however, provide a structure for exploring certain more extreme reactions to various standards proposed.

A workshop was held at IIASA[5] to explore methodological issues in standard-setting. Uncertainty of information, conflicts among actors and institutional constraints emerged as central to limiting the usefulness of modelling efforts. Table 2 presents a summary of the usefulness and limitations of the approaches used in our analyses. A standard-setting laboratory bringing together regulator, developer, impactees, experts and analysts to interact around a model would be of interest. Here the analyst's model would stimulate discussion as

---

[5] IIASA = International Institute for Applied Systems Analysis, Laxenburg, Austria.

TABLE 2

Uses and Limitations of Policy Analysis, and Decision Theoretic and Game Theoretic Approaches to Standard-Setting

	POLICY ANALYSIS	DECISION THEORETIC MODELS	GAME THEORETIC MODELS
USE	• to create a wide perspective of standard setting problems • to create problem structure, to identify groups involved, alternatives, and necessary information	• to structure decision problems • to quantify intangibles and uncertainties • to make trade-offs explicit • for conflict and sensitivity analyses	• to structure dynamic decision process • to explore parametrically the future of standard setting processes • to identify sensitive decision points
LIMITS	• mainly descriptive and qualitative • problem, not solution oriented • possible misperceptions	• static, limited feedbacks • problems of intergroup trade-offs • problems of quantifying political objectives	• highly abstract and aggregated utility functions and transition probabilities • arbitrariness of solution concept

RECOMMENDED USE IN STANDARD SETTING PROCEDURES	• in the **pre-standard setting phase** for:
	  — structuring problem
	  — ordering alternative perspectives
	  — specifying perceived roles and information
		• in the **standard setting phase**, after a clear decision problem has been formulated for:
		  — evaluating information
		  — quantification
		  — trade-offs and sensitivity analysis
			• in the **standard setting and post-standard setting phases** for:
			  — simulation of future changes
			  — sensitivity and conflict analyses

well as demonstrate the usefulness of the proposed model. Representatives from both Norway and the U.K. who attended our workshop demonstrated interest in the idea of a modelling laboratory as an aid to regulatory decision-making.

Conclusions

The case study on oil pollution in the North Sea from offshore platforms provided a rich source for studying the standard-setting problem. Primarily, it showed that the wide framework used was helpful in laying out the issues involved.[6] The actor structure was helpful for problem and option definition as well as for shaping the ensuing model analysis. Defining the problem in both Norway and the U.K. allowed an interesting comparison to further highlight the issues.

Systems analysis or synthesis was also a valuable perspective in that it showed the context of setting offshore discharge standards. For example, the severity of oil discharge standards offshore could affect oil discharges nearshore or onshore as well. If the offshore standard is low, say 0-20 ppm, and if it is coupled with a rigorous offshore monitoring, inspection and sanction programme, then discharges might be expected closer inshore or even onshore. Therefore, given a cost constraint to the company, a tightening of controls in one place might mean a loosening in another place where environmental effects may be greater such as in a biologically productive estuary. We found no evidence of a systems view in either country, which covered either the spectrum of oil operations or the spectrum such operations would affect. The regulatory authorities involved had rather accepted a purely pragmatic view from the beginning.

Institutionally, the study pointed out the importance of the procedural premises and sequence in standard-setting. Both uncertainty of information and actor value conflicts give great significance to the actual decision process. It appears that standard-setting can become more formal as the highly empirical approach adopted by regulators becomes more obvious. A more open consultative process can be anticipated along with a formal record of the opinions and criticisms of regulators, developers and experts. This greater accountability by technologists and their regulators, is to be welcomed in an age of increasing interdependency and threats to lifestyles.

References

Committee on Environmental Decision-Making (1977). "Decision-Making in the Environmental Protection Agency". National Academy of Sciences, Washington, D.C.

---

[6] This procedure was useful in both setting discharge standards and clarifying safety problems. The latter aspect is discussed in my forthcoming article and in IIASA publication RM78-6.

Cowell, E.B. (1977). Oil Pollution of the Sea. *In* "Marine Pollution," (R. Johnston, ed.). Academic Press, London.

Fischer, D. (1975a). On the problems of measuring environmental benefits and costs. *Social Science Information* **13**, No. 2, 95-105.

Fischer, D. (1975b). Willingness to pay as a behavioural criterion for environmental decision-making. *Journal of Environmental Management* **3**, 29-41.

Fischer, D. (1978). "A decision analysis of the oil blowout at bravo platform." RM 78-6, IIASA, Laxenburg, Austria. (also published in *Energy: An International Journal* **3**, 785-797.

Fischer, D. and Keith, R. (1977). Assessing the development decision-making process: A holistic framework. *American Journal of Economics and Sociology* **36**, 1-17 (January 1977).

Fischer, D. and von Winterfeldt, D. (1978). Setting standards for chronic oil discharges in the north sea. *Journal of Environmental Management* **6**, 177-199.

Holden, M. Jr. (1966). "Pollution Control as a Bargaining Process: An Essay on Regulatory Decision-Making". Cornell University Water Resource Center, Ithaca, New York.

Keith, R., Fischer, D., Francis, G., Lerner, S., De Ath, C., and Farkas, E. (1976). "Northern Development and Technology Assessment Systems: A study of Petroleum Development Programs". Background Study No. 34. Science Council of Canada, Ottawa, Canada.

Majone, G. (1975a). "On the Logic of Standard-Setting Health and Related Fields". IIASA Research Report, Laxenburg, Austria, (March 1975).

Majone, G. (1975b). "Standard-Setting and the Theory of Institutional Choice". IIASA Working Paper, Laxenburg, Austria (July 1975).

Pearce, D.W. (1976) "Environmental Economics" 96-103. Longman, London.

Majone. G. (1978). "Environmental Standard-Setting: Efficiency, Equity, and Procedural Problems". IIASA Mimeo, Laxenburg, Austria.

Schon, D. (1971). "Beyond the Stable State", 31-52, 245-246. W.W. Norton and Co., New York.

von Winterfeldt, D. (1978), "Modelling Standard-Setting Decisions: An Illustrative Application to Chronic Oil Discharges". RM78-27, IIASA, Laxenburg, Austria (May 1978).

"REAL" VERSUS PERCEIVED RISK: IMPLICATIONS FOR POLICY

Raphael G. Kasper

*Environmental Studies Board,
National Academy of Sciences,
Washington, D.C., U.S.A.*

Uncertainty pervades our lives. Modern technological society has become so complex, so difficult to comprehend, that virtually every activity we undertake carries with it consequences that are unknown, or at best unclear, and which we can only guess about or attempt to predict using feeble and imperfect tools. This state of affairs is made particularly distressing by the fact that the set of consequences that accompanies any activity may include potential threats to our surroundings, to our well being, or to our way of life.

Decisions about the course of society must somehow identify and evaluate the nature and extent of the potential threats and, to this end, research on methods of risk assessment has been vigorous and occasionally insightful and successful.

The evaluation of risks is, of course, only a part of any decision-making process. A rational decision-maker must also determine and evaluate a set of benefits of a given activity and then weigh the benefits against the costs and risks. The whole process is often intuitive rather than explicit; the point here is merely that risk evaluation (or risk assessment) is a piece and not the totality of the decision-making process. Despite all of the problems of risk assessment, it could be argued that it is not even the most difficult part of the process, but that is best left to other forums.

Another word about risk: many discussions of risk and risk assessment consider only threats to human health and life. In fact, technology and other activities of people and societies can pose other threats, threats to animals, plants, forests, and oceans and threats to continuation of styles of life and enjoyment of life. Considerations of such risks must be a feature of technological decision-making lest our attempts to limit the effects on health of certain activities create longer term problems that may weigh every bit as heavily on humanity.

The rapidly growing literature in the field of risk assessment techniques draws a distinction between two measures of risk that can be called, for want of better terms, objective and subjective. Objective measures are said to describe the "real" or actual risk of a process or project. They seek to tell us, for example, what the chances are of an explosion at a liquid natural-gas terminal, or what the probability is of an oil spill, or what the expected health effects are of various levels of atmospheric pollutants. Sometimes experimental evidence or long term experience provides useful information with which to estimate the risk; sometimes sophisticated calculations can be used to project potential risk. Subjective measures rely upon the perceptions of those assessing the risk. They tell us what people think the risks of a particular activity are. Numerous studies have demonstrated that objective and subjective measurements of the risk of a given endeavour rarely, if ever, agree. The psychological reasons for the disparity are interesting and, sometimes, provide great insight into the way in which people react to technological undertakings. But the differences between the two measures of risk, beyond their intrinsic interest, lead to a number of policy implications that have affected and will continue to affect the course of technological progress.

There have been a number of attempts to estimate the actual risk of an activity, although if one such effort could be thought of as capturing the essence of what has been referred to as objective measurement of risk it would be the "Reactor Safety Study: An Assessment of Accident Risks in U.S. Commercial Nuclear Power Plants", often referred to as the Rasmussen Report (1975), after its chief architect. That study applied techniques developed in the U.S. space programme and elsewhere to an estimation of the probability of various occurrences in nuclear power plants and was presented, with much fanfare, as a demonstration of the safety of the nuclear power programme. Other, less ambitious, examples abound.

One attraction of objective measures of risk is the apparent elimination of subjective elements in at least one part of the decision process. It is probably wise to lay aside this presumed benefit of objective measures on at least two grounds. Even the objective measurement of risk must, as do all intellectual endeavours, involve some element of subjectivity. The very choice of questions to be asked, issues to be considered, and methods to be used involves judgement. Thus in examining, let us say, the risk of nuclear power plant accidents, why does the analyst choose to consider particular chains of events and not others? The choice is subjective and based, at least in part, on the analyst's judgement of what may be the most important modes of failure. In this light, it is difficult, or impossible, to conceive of any analysis that is purely objective.[1]   Any attempt to reduce risk assessment to a rote, mechanistic application of technique is doomed to failure (although I fear that we will not escape bombardment with schemes that purport to take judgemental elements out of analytical techniques or to completely separate facts and values).   But even if the attempt could succeed, it would not have the effect of eliminating subjective values from the decision-making process for, as noted earlier, that process has other components, most notably the weighing or balancing of risks and benefits, that invariably involve judgement.

I leave it to others to discuss in detail the mechanics of risk assessment;   it is, however, worth bearing in mind that even the most coldly analytical of methods is likely to be inaccurate and potentially misleading.   The problems that bedevil the analyst are well known: paucity of carefully gathered data, difficulties of extrapolating certain effects from one population (often animal) to another (often human), uncertainties in the effects of chemicals or other pollutants in concert, unequal distribution of risks within the population.   The list goes on;   the important point is that the precise numbers that often result from analyses of risk carry with them a spurious appearance of great accuracy.

While technical experts and professional risk assessors struggle with the complexity of their craft, subjective measures of risk dominate the thinking and actions of most individuals.   The body of work by Paul Slovic and his colleagues (e.g. Slovic *et al.*, 1976) provides irrefutable evidence that perceptions of people about risk do not always (and, in fact, rarely) coincide with what we know to have been the actual risk of certain activities.   Slovic and others have provided hypotheses that seek to explain why people don't always see things the way they are.   The implications of the difference between the risks people think they confront and the risks they actually confront (or, more accurately, the risks that experts think they actually confront) are the focus of this discussion.

Most, though not all, of Slovic's examples illustrate discrepancies between "real" and perceived risks in which the nature of the risk (often death or injury) is well understood.   The discrepancy can often be traced to the difficulty that people have in dealing intuitively with probabilistic phenomena or to a lack of complete or accurate information about the frequency of occurrence of

---

[1]Scientists and other technical experts display an evident discomfort with subjective values and judgements.   There is an irony here though;   while there are incessant discussions among scientists of the importance of doing "objective science," in all cases it is the value system of the scientists that dictates their choice of problems and methods of attack.

certain events.

But the inability of individuals to comprehend and process statistical evidence is not the only cause of differences between "real" and perceived risks.   Two anecdotal examples of another kind of disparity, perhaps as important, between "real" and perceived risk may serve to provide some background for what follows.

About six years ago, standing on a beautiful hilltop on a farm in Vermont, the leader of a citizens group seeking to deny an operating licence to the Vermont Yankee Nuclear Power Station, was relaxing after another day in what was, to her, a frustrating and protracted battle against an intransigent bureaucracy.   A well-educated non-scientist, she had read assiduously about nuclear power and its problems and had a good appreciation of the technology. All of a sudden, in the midst of a discussion having nothing to do with nuclear power she stopped, looked down the hill toward the Connecticut River where the power plant was located, and asked, "Can you imagine a huge mushroom shaped cloud down there?"   She knew that the power plant could not explode like a nuclear bomb.   And yet, somewhere in her mind, near enough to the surface that it could emerge without prodding to interrupt an otherwise peaceful moment, was a deep and pervasive fear of the power plant.   No number of Rasmussen reports or critiques of them could change the way she, and one must suppose, many others felt about nuclear power.

Near Santa Barbara, California, plans have been made to construct a large liquid natural-gas terminal.   The site was chosen by utility companies but was also recommended by the staff of the California Public Utilities Commission.   A small group of American Indians is protesting the siting of the plant, contending that the chosen location is the most sacred spot of their culture, the Western Gate, where the Indians say the souls of their people must pass after they die to join the spirits of their ancestors. Seeking to maintain the traditions and beliefs of a culture already ravaged by modern civilization, the Indians fear that the proposed terminal would deny the souls of the dead a passageway to the Gate.   One can envision a Rasmussen-like report, examining the probability that a soul would be unable to migrate past a liquified natural-gas terminal. The report would, no doubt, conclude that the risk was very small, perhaps 10^{-9} for each soul.   To technical experts the risk seem negligible;   to the Indians the risk is unthinkable, striking at the very core of their beliefs and, indeed, of their existence.

The importance for policy and decision-making of the difference between objective and subjective measurements of risk depends, to some extent, upon the type of society within which the decisions will be made.   It seems clear that the implications of the disparity are most important in a democratic society and less important in an autocratic or totalitarian society (or at least in a society that treats its decisions about technological undertakings in an autocratic manner).[2] But in any society, the distinction between the two kinds of risk creates difficulties for decision-makers and regulators.

The root of the problem seems to lie in the tendency of technological experts and many government decision-makers to view objective characterizations of risk as somehow more real or more valid than perceptions of risk. For most people, however, it is the perception of risk that has the most immediate effect on their thoughts and actions while objective characterizations seem abstract and unduly complicated.

One evident implication of the disparity has been an enormous emphasis on propaganda and indoctrination about technological issues, often at the expense of real progress toward solutions to existing or potential problems. In many areas, technical experts seem to have taken their mission as one of finding ways to manipulate public perceptions so that they more closely approximate the calculated or projected risk.   The history of the U.S. nuclear power safety programme, from its start through the Rasmussen report, is a good example of this tendency and is, as well, an example of the difficulty of manipulating perceptions of certain kinds of risk.   The drive to convince the public of the absence or inconsequence of risk is often carried to great, and some would say ludicrous, extremes.   The effort is usually futile.   Take for example, the attempt to convince airport neighbours that the Concorde SST is not a noisy airplane when a single overflight produces more information and data to the contrary than any number of newspaper advertisements and television announcements, filled with numbers and measurements, could ever provide. Similarly, a few more incidents like the liquified natural-gas truck disaster in Mexico July 1978 are likely to have a far greater effect on public perception than gas company publicity about the safety of LNG transport and storage.

If one could be confident that the public indoctrination programmes of industry (and often government) did not materially affect programmes designed to reduce risk to the public, there would be less cause for alarm. But what evidence there is on this issue argues to the contrary. Nuclear safety programmes and nuclear waste disposal programmes were clearly neglected through the 1960s and early 1970s (and, some would say, even in the present) as though the experts directing and designing the programmes were being swayed by their own public relations programmes. Recent developments (the Brown's Ferry fire and the discovery

---

[2] The author's experience is primarily with decisions in the United States, a society that, in dealing with technological issues, falls somewhere between the two extremes.

of potential difficulties in the emergency coolant containment
systems of certain boiling water reactors, for example) have
led to difficulties that might have been avoided had just a little
effort been shifted to solving problems and away from convincing
the public there were no problems.   Nuclear power is not the
only culprit;   inordinate public relations efforts have gone into
attempts by certain utility companies in the United States to
convince people that relatively inexpensive technology (tall stacks)
could protect the populace as well (from what the utilities characterize
as small risks in any case) as expensive technology (scrubbers),
and all the while development of scrubbers has proceeded at
a snail's pace. While their pollution control programmes languished
through the late 1960s and early 1970s, American automobile
companies invested large sums in advertising campaigns to convince
people that automobile emissions were not unhealthy.   It was
only the force of new laws that caused the manufacturers to
put real effort into their pollution control programmes. Despite
the prevalence of examples, it is difficult to think of a case in
which the indoctrination effort has succeeded. This is not surprising;
there is every reason to believe that, once divergent opinions
are established, only overwhelming and incontrovertible evidence
(impossible to obtain in many technological controversies) will
lead to consensus.

Perhaps the most telling effect of the disparity between expert
calculations of risk and people's perceptions of risk is the deepening
of the rift between technical experts and the rest of the public.
Distrust and misunderstanding are exacerbated.   When, as is
the case with many technological enterprises, the public sees
or senses larger risks than do the experts, the result is confusion
and disaffection. Technical experts begin to doubt the rationality
of the public and decry the emotionalism that stymies attempts
to encourage technological progress.   The rest of the public,
for its part, tends to see the experts as insensitive and, perhaps,
dangerous. The result is confrontation, often bitter, and occasionally
violent.   The indignant reaction of nuclear power experts to the
fact that the Rasmussen report was not accepted as a definitive
demonstration of the safety of nuclear power and the vigorous
protests of citizen groups at the site of the Seabrook, New Hampshire
nuclear power plant are examples of the two sides of the problem.

A related difficulty involves finding ways to admit the existence
of uncertain and incomplete knowledge in the presentation of
assessments of the consequences of actions with large technological
components.   Faced with possible adverse reactions from lay
citizens and government decision-makers, analysts are tempted
to overstate the accuracy and precision of their estimates of
risk (and often succumb to the temptation). But, as noted earlier,
even the best risk assessments are likely to be characterized
by large errors. It is quite understandable that the experts would

tend to profess more certainty than they actually possess; to do otherwise is to invite the experience of an environmental engineer who, when forced to admit that he was not absolutely certain that there would be no tritium release from a nuclear plant, faced hours of inane questioning about the effect that less than one part in 10^{23} of tritium would have on a river ecosystem. But the dilemna is two-pronged. The expert who claims excessive accuracy may be shown to be wrong; even if the point on which the expert is exposed is unimportant, credibility may be damaged to the extent that even accurate statements are rejected.

The difficulty of admitting uncertainty or absence of information has led to the feeling that if we lack any information it is necessary to obtain it. There has been a presumption, at least in recent years in the area of environmental risks, that all gaps in knowledge should be filled. That presumption is costly and often unproductive; some information may be of great intellectual value but have little bearing on decisions about protection of the public. Given limited resources for extending the frontiers of knowledge, it seems reasonable to place a higher priority on obtaining information that might have a significant effect on policy decisions. Yet, for example, relatively large sums of money have been spent to provide accurate models of air chemistry to relate automobile emissions to ambient atmospheric quality, models that are so much more accurate than our knowledge of the deleterious effects of the pollutants involved that the models can have no real effect on pollution control decisions.[3]

The discrepancy also makes more difficult the already vexing problem of establishing priorities for government action. Simply put, the question is whether to give precedence to problems for which the objective measures of risk indicate a problem or to pay most attention to those issues that disturb or frighten the public. Clearly, the position of the decision-maker relative to the involved public is important in this regard. The more directly the decision maker is accountable to the public, the greater the likelihood that public perceptions of risk will play an important role in the setting of priorities. This accounts for the fact that

---

[3]It is interesting to note that similar modelling efforts in a related but different area may be extremely critical to decision making. The decision is whether to use tall stacks to disperse sulphur oxide emissions from fossil fuelled power plants or flue gas desulphurization equipment (scrubbers) to minimize the emissions. Here the air chemistry of sulphur compounds, and particularly the rate at which sulphates are formed from sulphur dioxide, is a vital piece of information since studies of effects on health suggest that sulphates may be more implicated in adverse effects on human health than sulphur dioxide itself.

legislatures seem more often swayed by perceived risk than do government regulators whose positions are more effectively shielded from public accountability.

One disturbing aspect of the implications of the differences between "real" and perceived risk is that none of the implications seem to have a positive impact on the policy formation process. Positive impacts can be imagined, although they tend to be positive only to a portion of the populace and, in addition, are difficult to prove or even demonstrate. Thus one could argue that the differences provide a check to the trend toward technological determinism in modern societies; but this is only an advantage to those who believe there is such a trend and are disturbed by it. What is more, it is difficult to demonstrate that public perceptions have much effect, beyond slight delays, on whatever degree of technological determinism exists. Conversely, one could argue that the demonstration and exposition of "real" risks protects a vulnerable public from potential dangers and assures that limited resources are put to use as effectively as possible. This too is difficult to prove, and is only an advantage to those who see no benefit in relieving public apprehension when that apprehension seems ill-founded.

If the policy problems caused by the difference between "real" and perceived risk seem straightforward or self-evident, solutions to the problems are not. We can reject as utopian the hope that the two views of risk will ever coincide; the challenge becomes one of finding means of effective policy making in the face of the disparity.

Two extreme solutions seem immediately evident, though both are far from satisfactory. One, closest in fact to what is often done, is to insist on using only objective measures of risk in decision-making and ignoring public perceptions. One clear result of such an approach is public dissatisfaction with the decision-making process and fear of new technology; but to the extent that the decision-making apparatus can, and is willing to, ignore the public outcry this approach enables the most rapid development of new technology and, in the eyes of the experts, the most effective protection against risk. The other extreme approach accepts and is directed only by the perceived risk; it reacts only to public perceptions of risk, giving priority to those programmes that will minimize the perceived risk. Such an approach, while designed to quell public apprehension, can result in excessive conservatism in the development of technology and often in a misdirection of priorities away from programmes that would provide the greatest possible protection of the public.

It is to be hoped that there exists a middle ground that will ease the problems of policy formulation. Finding the middle ground is a challenge to government, to the public, to technical experts, and to the developers of new technologies. It would, or should, involve early and real involvement of all affected parties. This sounds eminently sensible but, in fact, it rarely happens. Most, if not all, early planning exercises, particularly those that involve energy facilities, have been charades (Ebbin and Kasper, 1974). Without a real inclination on the part of everyone involved to co-operate and to avoid conflict, future attempts are likely to be charades as well.

Such a general and vague conclusion may leave the reader uneasy. It should. It certainly leaves the author unsatisfied. Absent a compelling solution to the problems cited above, the lesson to be learned here is that problems of objective and perceived risk, interesting as they are as matters of purely intellectual and conceptual concern, have real and pervasive policy implications that have significant impacts on the way we think, act, and progress in modern society.

References

Ebbin, S. and Kasper, R. (1974). "Citizen Groups and the Nuclear Power Controversy". MIT Press, Cambridge, Massachusetts.

Rasmussen, L. (ed.). (1975). "Reactor Safety Study: An Assessment of Accident Risks in U.S. Commercial Nuclear Power Plants". Report, WASH-1400. U.S. Nuclear Regulatory Commission.

Slovic, P., Fischoff, B. and Lichtenstein, S. (1976). "Cognitive Processes and Societal Risk Taking". Oregon Institute Monograph.

PART 2
RISK DATA AND THEIR INTERPRETATION

INTRODUCTION

In the preceding Part the need for both general and specific data on risks and environmental impact was made evident. In this Part we explore the question of whether risks are increasing or not and then take up the problems of quantification and perception of risks. The first two papers view the problem of determining the level of technological risk from two different perspectives and come up with seemingly contradictory results. However, further consideration provides some insights, reducing the incompatibility.

The first paper, by Harriss, Hohenemser and Kates, discusses the problem of determining the full scope of technological hazards. Such questions as whether technological development increases or reduces both the risk burden and benefits to society are addressed, but only the risk side is examined in detail. The main relationships studied are those between categories of disease and indicators of technological development. While both acute and chronic disease are considered, the major effort is directed at determining the percentage of mortality that is preventable by adjustments in technology. This is difficult at best and any results are tenable only within limits. In addition, morbidity, environmental-impact costs, impacts on eco-systems, and the cost of risk and hazard management are considered as measures of importance. The authors also try to generate estimates from available data.

In spite of the paucity of risk data relating cause and effect (either causally or statistically), the authors conclude that 15-25 per cent of human mortality is a result of technological hazard, with losses of \$50-75 thousand million annually. As much as \$200-300 thousand million per year is consumed by the hazard burden, hazard management and coping with hazards. The principal result of such expenditure has been the elimination of acute effects related to infectious disease and point-source pollution, with little progress in the areas of chronic disease and ecosystem-impact. They further conclude that technological hazards on balance are not getting worse, but warn that only the assessable parts of the problem have been addressed; the final answer is not yet apparent.

The second paper, by Leonard Sagan and A.A. Afifi, also examines available data on mortality, morbidity, and average life expectancy, but from a different perspective, namely, what factors work towards decreasing morbidity and mortality and increasing lifespan. This paper provides a contrast to the approach which considers only factors that might increase risk. In this study the authors survey data from 150 countries containing 99 per cent of the world's population. Over twenty variables were investigated and correlations sought. The stage of economic development of a country is a major factor and the authors have used energy consumption *per caput* in equivalent kilograms of coal (C.E.) used per year: Phase 1 – undeveloped (0-99 kg C.E.), Phase 2 – initial development (100 - 1999 kg C.E.). Phase 3 – developing country (2000 - 3999 kg C.E.), and Phase 4 – industrialised country (over 4000 kg C.E.). These phases are used only as a measure of economic development, and most factors were analysed "cross-sectionally" and "longitudinally" as explained in the paper. However, results for life expectancy at birth and infant mortality showed significant correlations with state of development, measured by both energy consumption and degree of literacy. Developing countries (Phase 2 and 3) show a direct correlation between life expectancy and energy consumption. However, Phases 1 and 4 do not, although there is a difference of over thirty-five years in life expectancy at birth between these two phases.

A number of propositions supported by both these conclusions and by those of Harriss *et al.*, can be drawn.

1) In undeveloped countries (Phase 1) the risks are from natural hazards, including drought and famine. Concern may be high in the Government and technological centres in these countries, but little can be accomplished without help from outside the country.

2) Technology and availability of energy in developing countries (Phases 2 and 3) generally extend life expectancy. In these countries the benefits of technology outweigh the resulting burden of risks. Thus, there is little incentive for controlling these risks.

3) In industrialized countries (Phase 4) the balance of technological benefit and impact may well be as described by Harris *et al.*, i.e. the balance is about equal but could possibly tip either way.

In this way, the two sets of data taken by different approaches for different purposes can be reconciled to provide some important insights on where and how technology delivers benefits and when it may become burdensome.

Throughout these discussions the interchangeable use of terms is a problem for both readers and practitioners. Harris *et al.* use "risk assessment" and "hazard management" in the way most of the authors use "risk identification and estimation" and "risk evaluation".

The next paper, by Thedéen, uses this statistical approach for specific problems in risk identification and estimation. Thedéen examines the problems of acquiring data and making analyses of risk for two particularly difficult situations: long-term health data where cause and effect relationships are not easily discerned and low probability, high consequence events. He indicates three methods for addressing these problems: subjective or expert judgements, model analysis, and objective risk data. He then shows how a limit analysis, using objective statistical data can be used to evaluate model and judgemental approaches. The reader may consider that Thedéen puts too little emphasis on "ordinary" risks from air and water pollution and extrapolation of animal data to man, but this is more a problem of the author's focus rather than intentional underplay of these factors.

Actuarial data would appear to be the most objective of statistics. They are empirically collected to summarize the observed (historical) frequency of specific effect-causing events, such as deaths from transport-accidents, and so should be the easiest to interpret for risk-assessment. But problems arise when the event occurs with very low frequency or when the causal linkage between effect and event is highly speculative and remains undemonstrated because of long time-lags between the occurrence of the supposed causal-event and the onset of the resultant effect. In both cases, risk-estimates become highly unreliable.

The fourth paper, by Talbot Page, examines the limitations of actuarial methods when facing these "zero-infinity" dilemmas in dealing with rare or latent effects of catastrophic proportions. The difficulty of getting data to test predictions on low probability and delayed events makes the "accountability" of the risk assessors tenuous at best. One must be concerned with the "incentives" attributed to the assessors, since there is a wide range of latitude in "objective" estimates which cannot be tested. Greater accountability for such assessments can only be brought about by direct and indirect testing of predictions, and the author describes ways to implement such testing.

Page then shows that risk estimation is to some extent an adversary process, even for actuarial estimates based on considerable data. Profit maintenance and credibility are just two of the goals driving the assessors as "adversaries". He then goes on to illustrate the "credibility game", using Bayesian probability structures as an example. The conclusions drawn are that even "objective" estimates of risk do not necessarily result in accurate estimation of the actual risks and their magnitude.

The final paper in this Part is by Bohman and is intended as a brief commentary on the paper by Page. It describes the practical approach used by insurance companies for risk estimation, based on more normal actuarial data. In addition, Professor Bohman has briefly touched upon a means of putting probabilities of risk into perspective by comparison with the life expectancy of a 60-year old male Swede. This is a most useful concept, since the manner in which risk estimates are perceived by the public is dependent upon benchmarks and comparisons with other risks. Life expectancy at various ages may be one set of benchmarks, easily understood by the general public, against which comparisons can be made.

THE BURDEN OF TECHNOLOGICAL HAZARDS

Robert C. Harriss, Christoph Hohenemser, and Robert W. Kates

*Hazard Assessment Group, Clark University,
Worcester, Mass., U.S.A.*

Hazards are threats to humans and what they value: threats to life, to wellbeing, to material goods and to environment. Today, hazards originating in both nature and technology are of major concern in developing and industrial nations alike. Coping with hazards involves a wide range of adjustments, from learning to live with hazard, to sharing the burden of hazard, to controlling and preventing death, injury, property loss and damage to human and natural environments.

Particularly in the industrialized countries, major efforts are being made on two complementary fronts: risk assessment and hazard management. Risk assessment involves the identification of hazard, the allocation of cause, the estimation of the probability that harm will be experienced and the balancing of harm with benefit. Hazard management involves the choice of options to be used in control and reduction of hazardous occurrences. As practised in the industrialized world, hazard management also involves an immense and growing bureaucracy, a series of seemingly irresolvable political battles, and an interplay between science and values that often confounds rational discussion. A key question in this discussion is, "how safe is safe enough?"

Risk assessment and hazard management inevitably involve a combination of scientific understanding and social judgement as to which risks to accept, which to reduce (and by how much), and when to forego or limit the use of a technology or natural location. In practice, we marshal our science and make our judgements one at a time, addressing a specific hazard or class of hazards.

To offer needed perspective, this article provides an interim report on our best estimates of the scope of the technological hazard burden in the United States of America.

Natural and Technological Hazards

For the majority of the world's people, living in the rural areas of developing countries, the hazards of greatest concern are ancient ones, and are predominantly rooted in nature. These natural hazards most often threaten agriculture, food-supply and settlement, and constitute a major burden. For example, geophysical hazards (floods, drought, earthquakes and tropical cyclones) each year in the developing world account for an average of 250,000 deaths and $15 thousand million in damage and costs for prevention and mitigation. (Burton et al., 1978). This is equivalent to 2-3 per cent of the gross national product (GNP) of the affected countries. Losses from vermin, pests, and crop disease are widely regarded as a larger problem (Porter, 1976). They can amount to as much as 50 per cent of food crops (Pimentel, private communication). And infectious disease, though declining, typically accounts for 10-25 per cent of human mortality, mostly among the very young (WHO, 1976).

By contrast, for industrialized nations, natural hazards are a much smaller problem. In the United States, geophysical hazards cause less than 1000 fatalities per year, and property damage and costs for prevention and mitigation, of the order of 1 per cent of the GNP. Vermin, pests and crop disease while occasioning serious losses, are kept in bounds by pesticides and other techniques, and infectious disease accounts for less than 5 per cent of human mortality. However, in controlling natural hazards to this extent, the industrialized nations have not escaped unscathed.

In the place of the ancient hazards of flood, pestilence and disease, new and often unexpected hazards, predominantly rooted in technology, have grown. These hazards, in our estimation are now as large or larger in impact than the natural hazards that they have replaced. We illustrate this point in Table 1 (Kates, 1978). As concrete examples of the large cost of technological hazards and their management, consider that the United States currently spends $40.6 thousand million per year or 2.1 per cent of its GNP on cleaning up air, land and water pollution (Council on Environmental Quality, 1978); that the cost of automobile accidents is estimated to be $37 thousand million, or 1.9 per cent of the GNP (Faigin, 1976); and that the toll of death alone from technological hazards involves in our estimate (see below) 20-30 per cent of male and 10-20 per cent of female mortality, and a value of medical costs and lost productivity of $50-75 thousand million, or 2.5-3.7 per cent of the GNP. Overall, expenditure and losses on technological hazards may be as high as $200-300 thousand million, or 10-15 per cent of the GNP (Tuller, 1978).

Technological hazards are thus big business, which in its scope is comparable to only a few sectors of national effort, such as social welfare programmes, transportation and national defence. And the impacts of technological hazards go well beyond mortality. To illustrate, we show in Table 2 the various groups, sectors, and environments affected, along with the dimensions of consequences that we consider in our work on technological hazard assessment and management.

TABLE I

Comparative Hazard Sources in the U.S. and Developing Countries

| | Principal Causal Agent(a) | | | |
| | Natural(b) | | Technological(c) | |
	Social cost(d) (% of GNP)	Mortality (% of total)	Social Cost (% of GNP)	Mortality (% of total)
United States	2-4	3-5	5-15	15-25
Developing countries	15-40(e)	10-25	? (f)	? (f)

(a)Nature and technology are both implicated in most hazards. The division that is made here is by the principal causal agent, which, particularly for natural hazards, can usually be identified unambiguously.

(b)Consists of geophysical events (floods, drought, tropical cyclones, earthquakes and soil erosion); organisms that attack crops, forests, livestock; and bacteria and viruses which infect humans. In the U.S. the social costs of each of these three sources are roughly equal.

(c)Based on a broad definition of technological causation, as discussed in the text.

(d)Social costs include property damage, losses of productivity from illness, or death, and the costs of control adjustments for preventing damage, mitigating consequences, or sharing losses.

(e)Excludes estimates of productivity loss by illness, disablement or death.

(f)No systematic studies of technological hazards in developing countries are known to us, but we expect them to approach or exceed U.S. levels in heavily urbanized areas.

TABLE 2

Technological Hazard Impacts

HAZARD EXPOSURE RECEPTORS	DIMENSIONS OF CONSEQUENCES
1. HUMAN POPULATIONS Individuals, groups, cohorts	well-being (diminution, loss); morbidity (acute, chronic, transgenerational); mortality (acute, chronic, transgenerational);
2. ECONOMY Activities, institutions, production	individual and collective loss; cost of control adjustment; cost of mitigation.
3. SOCIETY Activities, institutions values	activity disruption; institutional breakdown; value erosion.
4. ECOSYSTEMS a. Biological	species extinction; productivity reduction; resistance/resilience diminution.
b. Environments (Natural)	landscape transformation; air and water quality deterioration; recreational opportunities lost.
c. Environments (Built)	community loss; architectural deterioration.

How may the full scope of technological hazards be evaluated? To answer this question involves the full sum of impacts and consequences as outlined in Table 2. This is a formidable job which no group has to our knowledge accomplished, or even attempted. In our work so far we have concentrated on human mortality and on ecosystems, particularly biological species and communities. The first is based on well defined data, and of all impacts, is most susceptible to quantification; the second is at best difficult to judge, and nearly impossible to quantify. They thus provide the range of current scientific understanding within which other impacts and consequences fall.

Human Mortality as a Measure of Technological Hazard

Human death is the best defined of all hazard consequences. Even many impoverished societies keep reasonable records, and for a large number of countries, including the U.S.A. (where the National Center for Health Statistics publishes vital statistics annually), mortality statistics, grouped according to "causes of death" are extensively tabulated by age, sex and even race (WHO, 1976). It would, therefore, seem that there should be a direct and obvious answer to the question, "how much death is due to technological hazard?" It seems simply necessary to add up the contributions of each "cause of death" and note the relative magnitude of the sum.

Unfortunately, once we have added the toll of transportation and occupational accidents and the impact of violence, this approach ends in a quagmire of uncertainty for at least three reasons:

1) death rarely has a single cause, and in most cases technology is at best a contributing factor;
2) when chronic disease, such as cancer or heart disease is given as "cause of death," we can deduce little directly about the role of technology, since the root causes of chronic disease are known in only a small percentage of cases;
3) much death is not accurately classified according to "cause," and this is true even of some cases involving accidents and violence.

Mortality statistics are further clouded by the fact that in many developing countries, as much as 50 per cent of all mortality is classified as of "unknown origin," and even in developed countries the practices in assigning cause vary widely (Preston, 1976).

In our study of technological hazards we circumvent these problems by using an indirect approach. Instead of estimating the percentage of mortality involved with technology by direct calculation and summing, we use a two-step process.

First we estimate the fraction of mortality that is preventable in principle, or equivalently involves external or non-genetic causes. In the literature this is often called exogenous mortality.[1]

Second, we estimate the percentage of technically preventable mortality that is involved with technology. In doing so, we recognize that externally caused or exogenous mortality sets an upper limit for technological causes of mortality, and in general contains social, cultural (sometimes referred to as life-style), and natural environmental components as well. The classification of exogenous

---

[1] The Latin "exogenous" means "of external origin." "Exogenous mortality" as used in our discussion should not be taken to deny all genetic involvement — it refers rather to that fraction of mortality which in a purely statistical sense can be altered by altering external conditions. Genetic factors, including inherited susceptibility to a particular disease, are by no means excluded. One need only regard disease in an individual case as a combination of genetic predisposition and external factors.

mortality (illustrated in Table 3) is not clear-cut. In an interrelated and mutually dependent society such as ours, most deaths have multiple causes.

TABLE 3

Classification of Morbidity and Mortality

	Cause	Examples
1.	Endogenous: causes reside predominantly within the individual	Ageing: genetic defects arising from inherited genetic load.
2.	Exogenous: causes reside predominantly outside the individual	
	a. Natural environmental	Infection; background-radiation induced cancer; latitudinal skin cancer effects; natural catastrophes
	b. Social and Cultural	Diet-based disease such as cancer from betel nuts, cirrhosis of liver from alcohol, heart disease from overweight; smoking related diseases; some urban related mortality; some violence; war-deaths.
	c. Technological 1. diffuse effects	pollution related diseases; some urban related mortality
	2. specific technology	transportation accidents; cancers from specific industrial chemicals such as benzene, asbestos, and vinyl chloride; gun accidents.

Exogenous Mortality

What fraction of mortality is exogenous, that is, preventable in principle? To answer this question, we first divide all of mortality into acute and chronic causes of death, shown in Tables 4 and 5. Among acute causes of death we include all those cases for which death is sudden and not preceded by a long period of illness. Among chronic causes of death we include all cases for which death results from a long period of prior morbidity or illness due to deterioration of one or more body functions. The division into acute and chronic causes is made because the analysis in the two cases is fundamentally different.

Acute Causes of Death Except for congenital malformations leading to sudden death, a small percentage of infectious disease, a percentage of accidents, suicides and homicides associated with inherited deficiencies and psychotic illness, all acute causes of death are *prima facie* exogenous. Assignment of the exogenous percentage is therefore made at or near 100 per cent in most cases, as shown in Table 6.

Chronic Causes of Death For chronic causes of death we obtain the exogenous pecentage by a comparison of the mortality statistics reported by 36 nations to the World Health Organization (WHO, 1976). The nations selected are believed to have sufficiently reliable statistics for our purposes; all have mortality rates for "unknown causes" amounting to less than 10 per cent of total mortality. From the 36-nation data, the lowest age-specific mortality rate was chosen, and taken as the "base case." Exogenous mortality for each nation was then *defined operationally* as the excess mortality observed in each relative to "base case" mortality.

There are several problems with this definition, all of which lead us to regard it as only an approximate estimate of true exogenous mortality. For example, the definition implicitly assumes that the genetic disposition of various populations toward mortality is identical. This is not always so. Some cancers, for instance, appear to have a genetic basis. On the other hand, when populations migrate they usually take on the mortality patterns of their new home, thus indicating the predominance of external factors. In addition, our method for obtaining the exogenous mortality depends critically on the validity of the "base case" as representative of near zero exogenous mortality. Our definition would tend to underestimate true exogenous mortality if some "base case" mortality is preventable in principle; and would tend to overestimate true exogenous mortality if the "base case" involves serious under-recording of certain chronic causes of death. Fortunately those latter effects, both of which are surely present, will at least partially cancel each other out.

To illustrate the kind of data that we have used, we show in Figure 1 age-specific cardiovascular and cancer mortality for males and females in selected countries, including the lowest and highest mortality cases. Male and female exogenous percentages deduced from this data are 80 and 60 per cent for cardiovascular disease, and 60 and 45 per cent for cancer, respectively. Exogenous percentages for all causes of death are summarized in Table 6.

TABLE 4

Acute Mortality in the United States in 1972 (WHO, 1976)

CAUSE OF DEATH	Mortality deaths/100,000		Mortality % of total (acute + chronic)	
	male	female	male	female
INFECTIOUS DISEASE	43.0	34.6	4.0	4.3
influenza	2.4	2.4		
pneumonia	31.9	23.7		
infection of the kidney	3.0	3.6		
enteritis	1.0	1.1		
infectious hepatitis	0.3	0.4		
other	4.4	3.4		
DEATHS IN EARLY INFANCY	27.2	19.6	2.5	2.4
diseases of early childhood	19.5	13.2		
congenital abnormalities	7.7	6.4		
TRANSPORTATION ACCIDENTS	43.1	15.7	4.0	1.9
automobile	39.6	15.1		
other	3.5	0.6		

TABLE 4 (Contd)

CAUSE OF DEATH	Mortality deaths/100,000		Mortality % of total (acute + chronic)	
	male	female	male	female
OTHER ACCIDENTS	_35.5_	_17.7_	3.3	2.2
poisoning	3.7	1.6		
falls	8.4	7.7		
fire	4.0	2.5		
drowning	5.0	1.0		
firearms	2.1	0.3		
industrial machinery	5.1	0.5		
others	7.2	4.1		
VIOLENCE	_32.9_	_10.5_	3.1	1.3
suicide	17.5	6.8		
homicide	15.4	3.7		
OTHER ACUTE CAUSES	_11.8_	_9.0_	1.1	1.1
TOTAL ACUTE CAUSES	_193.5_	_107.1_	18.0	13.2
MALE-FEMALE AVERAGE			15.6	

TABLE 5

Chronic Mortality in the United States in 1972 (WHO 1976)

CAUSE OF DEATH	Mortality deaths/100,000		Mortality % of total (acute + chronic)	
	male	female	male	female
CARDIOVASCULAR DISEASE	554.7	459.3	51.7	56.7
Hypertension	9.5	10.9		
Ischaemic heart disease	382.4	277.6		
Cerebrovascular disease	94.0	110.5		
Arteriosclerosis	29.2	26.7		
other cardiovascular	39.6	33.6		
CANCER	188.1	149.6	17.5	18.5
lung, trachea, bronchia	56.8	14.0		
colon	17.4	18.8		
breast	0.3	29.2		
lymphatic tissues	10.5	8.4		
prostate	18.0	–		
stomach	9.2	5.8		
leukemia	8.1	5.8		
uterus	–	6.0		
rectum	5.6	4.2		
mouth–pharynx	5.3	2.0		
other	56.9	55.4		

TABLE 5 (Contd)

CAUSE OF DEATH	Mortality deaths/100,000		Mortality % of total (acute + chronic)	
	male	female	male	female
CHRONIC LIVER DISEASE	<u>36.7</u>	<u>31.8</u>	3.4	3.9
diabetes	15.6	21.4		
'cirrhosis	21.1	10.4		
CHRONIC RESPIRATORY DISEASE	<u>25.8</u>	<u>7.6</u>	2.4	0.9
tuberculosis	2.5	0.9		
bronchitis, emphysema, asthma	23.3	6.7		
OTHER CHRONIC DISEASE	<u>74.6</u>	<u>55.1</u>	7.0	6.6
TOTAL CHRONIC DISEASE	<u>879.9</u>	<u>703.4</u>	82.0	86.8
MALE-FEMALE AVERAGE			84.4	

TABLE 6

Estimated Exogenous and Technologically Involved Deaths in the U.S.A.

CAUSE OF DEATH	Estimated exogenous component of mortality %		Estimated technological component of mortality %		(annual deaths in thousands)	
	male	female	male	female	male	female
ACUTE MORTALITY						
Infectious disease(a)	90	90	0	0	0	0
Deaths of infancy(b)	50	50	5	5	1	1
Transportation accidents(c)	100	100	90	90	90	39
Other accidents(d)	100	100	70	50	28	11
Violence(e)	100	100	30	30	10	3
Other acute deaths(f)	100	100	70	50	8	5
CHRONIC MORTALITY						
Cardiovascular disease(g)	80	60	0-40	0-40	0-217	0-132
Cancer(h)	60	45	40	25	82	35
Chronic liver disease(i)	80	80	0	0	0	0
Chronic respiratory disease(j)	60	10	0-20	0-5	0-5	0
Other chronic disease(k)	70	70	25	25	19	15
ALL MORTALITY			18-30	11-21	193-322	88-170

TABLE 6 (Contd)

(a) Exogenous percentage of 90 per cent based on the hypothesis that this amount of infectious disease is in principle preventable before genetic factors become dominant. Supportive of the hypothesis is that the declining trend of infectious disease mortality is steep. The technological percentage of zero is based on the fact that infectious disease is usually prevented by technology, not enhanced.

(b) Presently, the U.S.A. ranks 17th in infant mortality, and even in the lowest nations, infant mortality is still declining. The estimate for the exogenous percentage is meant to reflect these facts qualitatively. The technological percentage is low because infant deaths result largely from disease.

(c) Transportation accidents are *prima facie* 100 per cent externally caused. The technological percentage given includes all deaths except those that are estimated to be predominantly homicidal or suicidal.

(d) Other accidents include numerous categories, as shown in Table 4. All are by definition externally caused. Some, like drowning and falls are primarily rooted in culture and society, not technology, and hence these are excluded in estimating the technological percentage.

(e) Although nearly all violence is committed with the help of technological devices, and thus suggests 100% exogenous causation, there is little evidence that violence is prevented by modification of technology. Rather, violence is rooted in culture and society. The assignment of a modest technological percentage reflects this fact.

(f) Other acute deaths involve many causes, but relatively small numbers. The values given represent the average behaviour of other acute deaths.

(g) The exogenous percentage is based on 36-nation comparisons as illustrated in Figure 1. The technological percentage is uncertain, yielding 40% based on cross-national plots similar to Figure 2, the difference between the U.S. rate and some theoretical rate without technology (0%), yet yielding near zero based on

(contd overleaf)

TABLE 6 (Contd)

state-by-state comparison within the U.S. similar to Figure 3. Since much of cardiovascular epidemiology points toward diet, tobacco, and stress, we are inclined to believe the lower technological pecentage.

(h)The exogenous percentage is based on Figure 1. The technological percentage Figure 2, is the difference between the U.S. rate and some theoretical rate without technology (0%), and the support given this by Figure 3 as well as the available literature on cancer epidemiology.

(i) The exogenous percentage is based on data similar to Figure 1. The low technological percentage is based on the predominant role of diet and alchol in liver disease epidemiology.

(j)The exogenous percentage is based on data similar to Figure 1. The technological percentage is based on the literature describing the urban-rural difference in epidemiological studies of smoking related disease.

(k)Other chronic diseases involve a large variety of causes, but rather small total mortality. Percentages assigned here are guesses, based on the average behaviour of the chronic diseases which we have analysed.

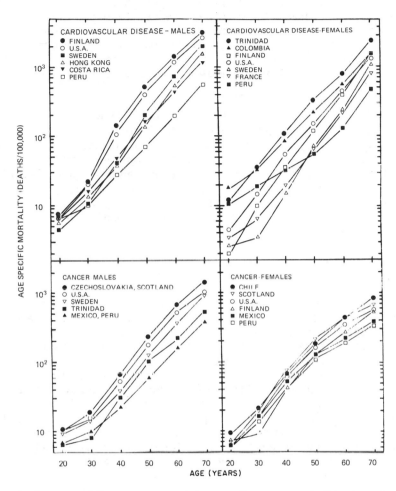

Fig. 1 Age-specific cancer and heart disease mortality in selected countries, for males and females during 1972-73. These countries were selected because they are believed to have reliable statistics and because they represent the full range of recorded mortality, from lowest to highest. The differences between the lowest and those for the U.S. were used to estimate the exogenous fraction of mortality for the U.S., as given in Table 6. Note that the plots shown here utilize a logarithmic scale. (WHO, 1976).

Technological Mortality

What percentage of exogenous mortality derives primarily from technology, as opposed to the natural environment, society or culture? This is a much more difficult question, with a considerably more uncertain answer than in the case of exogenous mortality *per se*. There is no simple argument that allows approximate separation of the technological percentage. Our present best estimate is thus pretty much a guess, though hopefully a good one. To proceed with this guess, we again treat acute and chronic causes of death separtely.

Acute Mortality   Infectious diseases, though influenced by the level of technology, is largely environmental and cultural in orgin. Technology usually leads to a reduction of disease, rather than increased hazard. In contrast, accidents, homicide and suicide are heavily involved with technology and culture, and marginally with the natural environment. Our estimate of the technologically involved percentage of acute mortality thus ranges from 0 per cent in the case of infectious disease to 90 per cent in the case of transportation accidents (See Table 6).

Chronic Mortality   We have already noted in our discussion of exogenous mortality that direct assignment of cause in the case of chronic disease is usually not possible. For estimating the technological component of exogenous mortality we again use an indirect method, based on national and international comparisons. Our approach is to look for correlations of chronic disease mortality with certain indicators of technology, such as *per caput* GNP, *per caput* energy consumption, and per cent of labour in manufacturing. If chronic disease increases with level of technology, this analysis yields the equivalent of a "dose-effect" relation: i.e. it permits the determination of the change in mortality with a given change in level of technology. Unlike the high quality dose-effect relations in the field of toxic substance epidemiology, the exposed populations in this case are poorly controlled for factors other than level of technology. Hence, one must expect a certain amount of scatter in mortality at a given level of technology.

The Case of Cancer   We illustrate our analysis for the case of cancer. International "dose-effect" relations for men and women are shown in Figure 2; equivalent relations for the U.S., for both blacks and whites, are shown in Figure 3. In each case the mortality in 1972-73 is plotted against per cent labour in manufacturing in 1940, thus allowing for the latency of cancer. Our interpretation of the observed relations are as follows:-

1) Internationally, cancer in males varies widely, and shows an average increase of a factor of 2.7 for males and 1.7 for females as the level of technology varies from lowest to highest. Particularly for males, the scatter is very large, indicating that there are many other causes at work.

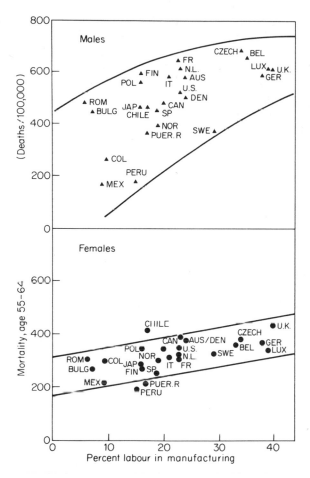

Fig. 2 Correlation between age-specific 1972-73 cancer mortality (WHO, 1976) and per cent of labour in manufacturing in 1940 (Woytinsky and Woytinsky, 1953) for nations believed to have reliable mortality statistics. Though the data exhibit wide scatter among nations, both males and females show increasing cancer mortality with increasing industrialization. The scatter indicates that there are causes other than industrialization. The consistent increase of cancer mortality with industrialization indicates that the latter is probably one of the causes of cancer (see footnote[2] p.121 on cause and correlation). The choice of 1940 to measure the level of industrialization allows for the known lag of approximately 30 years between exposure to carcinogens and the occurrence of cancer. Note that, consistent with their greater participation in industry, males show a bigger increase than females. These data were used to estimate the fraction of technologically involved mortality given in Table 6.

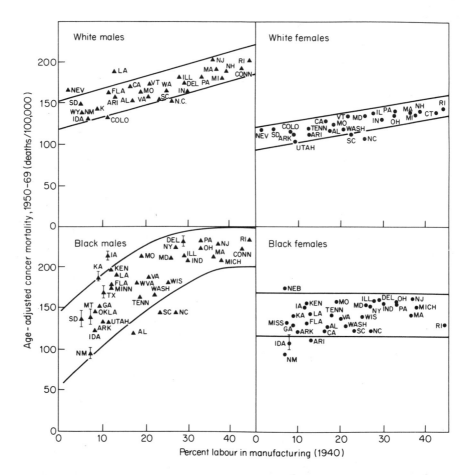

Fig. 3 Correlation between average 1950-1969 age adjusted cancer mortality and per cent of labour in manufacturing in 1940 for states within the United States. As might be expected from the greater homogeneity of the U.S., the scatter is considerably smaller than that for the international data shown in Figure 2. The pattern of increasing mortality with increasing industrial exposure is repeated. Again men show a more pronounced increase than women, and in addition, black men show a bigger increase than white men. The only surprising aspect of the data is that black females show no apparent increase. (National Institute of Health, 1970).

2) Within the U.S. for whites, the international pattern is repeated, though with smaller increases, and less scatter. Thus mortality increases by an average factor of 1.5 for males and 1.4 for females as

level of technology ranges from lowest to highest.

3) Within the U.S. for blacks, the pattern is significantly different. For males the increase in mortality is on average a factor of 2.0, as technology varies from lowest to highest. Compared with whites, this is a distinctly larger effect. For females, on the other hand, no distinct effect is seen, though scatter is large, and average values are higher than those for white females.

Thus, although there are some puzzles, a reasonably consistent picture emerges. Cancer, as one would expect from the epidemiological literature, (Fraumeni, 1975) has an appreciable technologically involved component of mortality. Using the international data shown in Figure 2, we estimate the difference between the U.S. rate and some theoretical rate without technology (0 per cent labour in manufacturing). This gives a conservative estimate of at least 40 per cent for men and 25 per cent for women in the U.S. Thus, very roughly speaking, about half of exogenous cancer mortality is involved with technology, the rest with social and cultural causes. Similar results are obtained if *per caput* energy consumption is used as an indicator of technology.

Does Technology Cause Cancer? We do not wish to claim that energy consumption or industrial employment causes cancer *per se*. Correlations such as shown in Figures 2 and 3 are too weak a tool for this purpose [2] However, when correlations with mortality emerge for

---

[2] To observe a correlation is not to establish a cause. A correlation is simply the regular change in one variable *associated* with a corresponding change in another variable. On a graph, cause and correlation look the same. How can they be distinguished? The best answer seems to be that to establish cause one tries to establish as close a link as possible, using when available, generally established theory and/or experimental evidence. To illustrate, consider three examples:

1) To-day most scientists are willing to agree that the force of gravity *causes* the earth to orbit around the sun, the tree to fall in the forest, and the tide to flow and ebb. Yet not so long ago, before an adequate universal theory of gravitation was formulated and applied, these three events were viewed as disparate phenomena, each understandable only in terms of a series of *ad hoc* assumptions, most of which turned out to be incorrect.

2) A more difficult case is the correlation between cigarette smoking and lung cancer. Though many scientists are now willing to say "smoking causes lung cancer," this was for a long time a disputed case because no general theory of cancer was and is available. Recently however, many of the substances in cigarettes and cigarette smoke have been isolated and used separately to expose animals experimentally, with the result that lung cancer developed.

(footnote contd overleaf)

several indicators across a wide range of populations and cultures, it is likely that the results are not accidental, but are evidence of a number of factors that form links in the causal chains directly or indirectly leading to observed chronic disease mortality. Sometimes these links are fairly simple and well-established: e.g. coal mining leads to a deposit of fine coal dust in deep-lung cavities, and through obstruction of these, reduces lung function (black lung disease). In other cases, the links are very complex, such as the incompletely understood connections between diet and heart disease. It is the task of medical science, particularly epidemiology, to identify and describe these specific links; and it is the task of hazard management to control them. Our purpose here is to explore the possible magnitude of the problem, and for this our correlations of disease with general indicators of technology are adequate and appropriate.

Using a method similar to the case of cancer, illustrated above, we have estimated the technologically involved percentage of mortality for other chronic disease as summarized in Table 6. We stress that by technologically involved percentage we mean mortality which is in principle preventable by adjustments in technology. This does not exclude other causative factors in this component of mortality, such as genetics, cultural milieu, life style and natural environmental conditions as *contributing* causes. To compensate, and to be conservative, we exclude smoking and diet as technological causes, even though technologies have figured highly in the consumption of cigarettes or in the availability of low-cost meat and dairy products.

The Cost of Technological Hazard Mortality

Estimates of mortality and morbidity costs for various causes of

---

(footnote contd)

3) With cancer of the colon there is a strong correlation with *per caput* meat consumption. Here the causal situation is highly ambiguous. The "cause" may be the meat itself; but it may also be the absence of high grain consumption which is normally present in cases of low meat consumption; and finally, it may be neither meat nor grain *per se*, but the way the meat is cooked or some other unrecognized factor. One must conclude, therefore, that here, turning correlation into cause is in its very early stages, and that by taking action one runs a fairly high risk of pursuing an irrelevant goal.

The specific links that buttress the involvement of technology with chronic disease span the full range of "strong" to "weak" causation illustrated by the above three examples. The overall correlation of chronic disease with technological indicators is itself closer to the weak end of the spectrum, but is probably the best that can be done at the present time.

death are available in the literature (Rice *et al.*, 1976).   These indicate the dollar value of medical care and the cost of lost productivity.   Such estimates do not place a dollar value on life and suffering *per se*, since this depends necessarily on many personal and societal ethical judgements that are widely held to be beyond economic valuation.   At the same time, such estimates are important because they define the magnitude of the economic problem of lost life and illness, and in this way serve to indicate the savings that may be realized if mortality and illness are avoided and prevented.

With the values for technology-connected deaths given in Table 6 and the estimated values of life-shortening applicable to each cause of death, we find that the total annual loss due to mortality from technological hazards is approximately 0.7-1.0 million person-years, about two-thirds of which applies to males.   Using the methodology developed by Rice, Feldman and White (1976) to translate this into medical costs and costs for lost productivity, we find an annual loss of $50-75 thousand million due to technology-connected mortality and related morbidity.   (Details of our analysis can be found in Goble *et al.*, 1978).   Interestingly, accidents and violence, though they constitute only 10.5 per cent of male and 6.5 per cent of female mortality, respectively, account for 40-50 per cent of costs.   This is because these cases generally have higher technological involvement and larger life-shortening effects.

Technological Impacts on Ecosystems

In contrast to human mortality, the ecosystem impacts on biological communities, while perhaps the most important of all dimensions of technological hazard in the long run, are also the most difficult to quantify.   Here there are no world-wide, nearly all-inclusive accounting systems such as death certificates.   And instead of dealing with a single dominant species, we are dealing with literally millions of species related by a complex and often fragile system of inter-dependence.   How may the impacts of technology on this system be defined?   Two measures of ecosystem impacts by technological hazards (see Table 2) are species extinction and ecosystem productivity.   Both of these measures are in principle quantifiable. Yet each has less specific meaning to humans and what they value than human mortality.   We consider each separately below.

Species Extinction

Species extinction is the most drastic and inclusive form of wildlife mortality.   Like human mortality it can occur naturally, independent of technological effects.   In analogy to human mortality, we are interested here in the percentage of species extinction that is of technological origin.   As before we divide the problem by asking two questions:

1) What is the rate of *exogenous* species extinction consisting of

cases for which the underlying causes are not of predominantly natural origin?

2) What is the rate of technologically involved extinction, consisting of the percentage of exogenous extinction that is predominantly related to technological causes.

One approach to the first question is through the historical record. As seen in Figure 4, this shows that the world-wide rate of

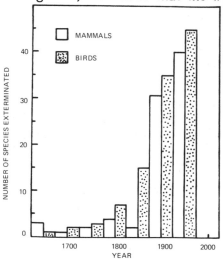

Fig. 4 The number of species of mammal and birds eliminated over the last three hundred years. Each bar represents a 50-year period. (Brunnel and Brunnel, 1976).

vertebrate extinction has considerably speeded up in the last 100 years, culminating in a current rate that is at least ten times the "baseline" or evolutionary rate observed 300 years ago (Ehrlich *et al.*, 1977). As shown in Table 7, in the U.S.A., one in ten species of native, higher plants is currently endangered, threatened with becoming endangered, or recently extinct; and in Hawaii, nearly half of the total diversity of the native vegetation is similarly involved (Brunnel and Brunnel, 1967; Fisher *et al.*, 1969; Vetz and Johnson, 1974; Council of Environmental Quality, 1975).

Another approach to estimating the exogenous extinction is through direct classification of species extinction according to cause. Using available data (Ehrlich *et al.*, 1977) on extinction and rarity of birds and mammals since 1800, we have obtained the division into exogenous and natural causes shown in Table 8. Thus for the period studied, more than two thirds of extinction and rarity has specifically non-natural causes.

How much of exogenous extinction is of distinctly technological origin? This question is unfortunately unanswerable in terms of well

TABLE 7

*Endangered, Threatened and Extinct Species of Native Higher Plants in the U.S.A.**

Status	Continental United States (including Alaska) Species, sub-species and varieties	%	Hawaii species, sub-species and varieties	%
Total native higher plants	20,000	100.0	2,200	100.0
Endangered§	761	3.8	639	29.0
Threatened¶	1,238	6.1	194	8.8
Extinct†	100	0.5	255	11.6
Total	2,099	10.4	1,088	48.9

*Source: Smithsonian Institution, (1975).

§"Endangered" is defined as in danger of becoming extinct throughout all or significant portion of their natural range.

¶"Threatened" is defined as likely to become endangered in the foreseeable future.

†"Extinct" is defined as limited to recently (or possibly) extinct species only: they cannot be found after repeated searches in the localities where they were formerly observed or other likely places. Some of the latter appear to be extinct in the wild, but are still preserved in cultivation.

R.C. Harriss et al.

TABLE 8

Classification of Causes of Extinction and Rarity for Birds and Mammals Since 1800 on a Worldwide Scale

Cause of extinction	Birds (%)	Mammals (%)
NATURAL CAUSES	24	25
EXOGENOUS CAUSES		
1. Acute (hunting)	42	33
2. Chronic		
a. habitat disruption (physical)	15	19
b. habitat modification (biological and chemical)	19	23
TOTAL	100	100

Cause of rarity	Birds (%)	Mammals (%)
NATURAL CAUSES	32	14
EXOGENOUS CAUSES		
1. Acute (hunting)	24	43
2. Chronic		
a. habitat disruption (physical)	30	29
b. habitat modification (biological)	14	14
TOTAL	100	100

Source: Recalculated from Fisher *et al.*, (1969).

defined analytical approaches. Technology is certainly heavily involved in hunting, and in much of physical habitat modification but we do not have the data for a case-by-case review of recorded extinction. In the absence of such detailed data, we conservatively place the technological percentage of exogenous species extinction at approximately one half, with the remainder largely of cultural character.

Whatever the division between technological and cultural may be, it is clear that the rates of exogenous extinction currently being observed are much faster than the normal evolutionary process of replacement. Nor is it possible to insure adequately against such loss in zoos, botanical gardens, and other protected environments (Ehrlich et al., 1977). Ecological theory, furthermore, suggests that wildlife mortality of the magnitude being currently observed can lead to significant diminution and loss of ecosystem productivity and resilience, with occasionally catastrophic consequences.

We wish to emphasize that counting species is inadequate by itself for defining the impact of technology on ecosystems. To predict the outcome of an evolutionary play, it is not enough to have a catalogue of characters. What is needed is some measure of the effectiveness with which ecosystems use energy, and how well an ecosystem is able to recover from a stress condition (resilience). Important new concepts related to ecosystem energy analysis (Odum and Odum, 1976) and ecosystem resilience (Fiering and Holling, 1974) are currently undergoing intensive study by the scientific community. Until these provide well-defined indicators, however, it seems prudent to use crude indicators like species extinction as warning signals of potential hazard.

Ecosystem Productivity

As a second measure of ecosystem impacts we consider productivity, or the ability of ecosystems to produce organic material from inorganic substrate and sunlight. In so doing, we limit ourselves to the changing magnitude of the land biomass – the organic material found on land. Land-biomass is subject to natural fluctuations from such factors as weather and disease; it is also affected by forestry, agriculture, urbanization, and similar pressures from humans. Biomass impacts can therefore, as before, be divided into natural and exogenous effects.

Global changes in land biomass have recently been explored in connextion with studies of the world carbon cycle (Bolin 1977; Woodwell et al., 1978). These studies show, albeit with great uncertainty, a net annual decline of 0.2 - 2 per cent in global land biomass (See Table 9). The causes of change are largely exogenous, and in areas of maximum population-pressure involve decline and destruction of major land-plant communities. Among the communities destroyed, tropical forests are of particular concern because it

R.C. Harriss et al.

is not clear that reforestation can take place on some lateritic soils. A detailed study of tropical forests indicates that about 0.3-0.6 per cent of the total is being destroyed each year (Sommer, 1976).

TABLE 9

Estimates of Current Net Loss of Major Land Plant Biomass, as Reflected by the Release of Carbon Into the Atmosphere

Plant community	Carbon released Average	1000,000,000 tons/yr range
Tropical forests	3.5	1-7
Temperate forests	1.4	0.5-3
Boreal forests	0.8	0-2
Other vegetation	0.2	0-1
Detritus and humus	2.0	0.5-5

Source: Modified from data given in Woodwell *et al.*, (1978).

In addition to direct losses in ecosystem productivity from deforestation, indirect impacts on drainage basins, resulting from major changes in hydrological and chemical cycles, can also diminish long term productivity of the total ecosystem. For example, replacing biomass and nutrients lost in harvesting northern hardwoods may take 60 to 80 years (Likens *et al.*, 1978).

As with the case of species extinction, exogenous decline of land biomass is of specifically technological as well as cultural origin. Because the bulk of the large changes now being seen, particularly in tropical forests, involve the application of high technology, we believe the technological component of biomass decline to be about 75 per cent of the total.

Technological Hazards in Historical Perspective

Our discussion so far has focussed on the impacts of present technological hazard. Except in the case of species extinction, we have made no effort to look at the historical record. Industrial development in the West is now 300-400 years old, and much of what has occurred in the past 50 years has been termed "post-industrial." Historical experience with technology is therefore extensive, and it is therefore interesting to ask "is the problem of technological hazards getting worse?" We discuss human mortality and ecosystem impacts separately.

Human Mortality

In regard to human mortality, the benefits of technology appear to have been large and dramatic. As already noted, they include the near elimination of the worst of natural hazards – infectious disease. This development is largely responsible for the fact that since 1850, when the U.S.A. had a highly dispersed agricultural population, life expectancy has shown a near doubling at birth, a 30-50 per cent increase at midlife, and a modest increase at age 60. Technology has also led to a food supply system that is so productive that few in the industrialized world need fear even slight deprivation of this basic need.

In addition, hazards of technology were undoubtedly higher in earlier, less fully managed stages of industrial development. Thus, occupational mortality at least of the acute variety, has shown a continuing and steady decline, as shown in Figure 5; and large technological disasters apparently peaked during 1900-1925 (National Safety Council, 1977). If evidence from literature is desired one need only recall the novels of Charles Dickens and D.H. Lawrence, which contain accounts of industrial pollution and human exploitation in an industrial setting that find few parallels in the modern age.

Thus, at worst the present problem may be that the positive effects of technology have for some time reached their maximum effect on human mortality, while the hazards of technology continue partially unchecked, affecting particularly the chronic causes of death that currently account for 85 per cent of mortality in the U.S.A. Supportive of this view is the fact that male life expectancy has not increased since 1950 and has even shown a slight decline.

But this view may be too pessimistic. Even the apparent increase in chronic disease, which forms the principal evidence for unchecked technological hazard mortality, may be erroneously interpreted. Thus, as shown in Figure 6, the age adjusted mortality from heart disease is declining along with most other causes of death, and increasing cancer mortality can in large part be explained by the delayed effect of earlier increases in smoking. In addition, there is indirect evidence (Preston, 1976) that certain chronic diseases were seriously under-reported in earlier parts of this century. Therefore,

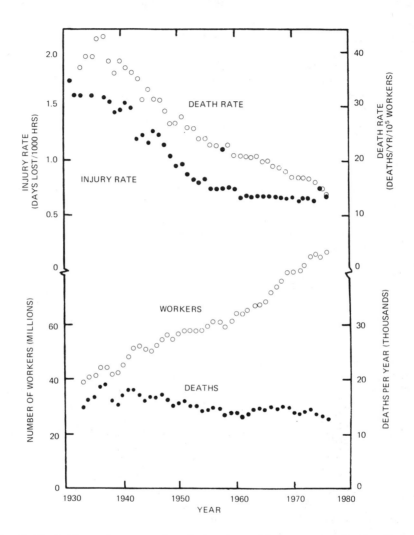

Fig. 5 Variation of occupational death and injury rates in the United States during the period 1930-76. Injury data are considered to be only approximate, since recording practices vary. (National Safety Council, 1977).

the actual mortality rates for cancer and heart disease shown from 1900-1940 were probably higher, and the overall increase since 1900 lower than shown in Figure 6.

In summary, we believe the burden of technological hazard mortality is not currently rising. Rather, it is clear that the last

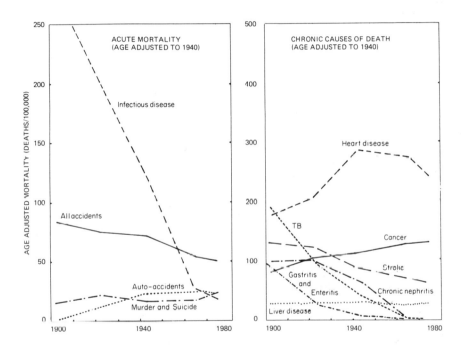

Fig. 6 Variation of age-adjusted causes of death in the United States from 1900 to the present. Among acute causes of death, note the sharp decline of infectious disease and the rise in automobile accident/mortality; among chronic causes of death, note the decline of most causes except for cancer and cardiovascular disease. Even the latter has declined since 1940. (Spiegelmann and Erhardt, 1974).

century in the U.S. has brought three things:

1) longer life through elimination of old ways of death which were largely acute and rooted in natural hazards;

2) an increase in chronic causes of death, rooted in part in technology; providing therefore;
3) a continuing burden of technologically related death, close to half arising from acciden⁺s and violence and the remainder from various chronic diseases.

Ecosystem Impacts

Beyond species extinction and productivity decline, what are the long-term trends in technological hazard impacts on ecosystems? We restrict ourselves to a few brief observations on what amounts to an enormous topic.

On the positive side, it is clear that massive, local releases of pollutants to the environment, as exemplified by the London killer smog, Minamata disease, and fish-killing concentrations of pesticides in rivers, are now less frequent. Trends in air and water quality indicate that after massive investments, environmental quality in the heavily populated and industrialized areas of the U.S. is generally improving (Environmental Protection Agency, 1977a,b). Thus strong control programmes for particulates and sulphur dioxide have reduced emissions to the point that very few urban regions are now experiencing violations of standards for these pollutants. Fish have returned to western Connecticut's Naugatuck River even in areas where no aquatic life could survive in the 1950s (Council on Environmental Quality, 1978). Interestingly, almost all of the major ecological hazards which have been identified and brought under control share two common attributes which determine the nature of the hazard management process – they originate from an early identifiable point source and are amenable to control by technological fixes of the source.

On the negative side of the ledger, it is equally clear that widespread release of pollutants in relatively low concentrations is degrading aquatic and terrestrial ecosystems at an unabated or even increasing rate. Calculated ratios of man-made to natural fluxes of heavy metals, for example, indicate that natural cycles of mercury, lead, antimony and selenium are being significantly altered by human activities (Stumm, 1977). The input of mercury to the global atmosphere from industrial and fossil fuel emissions exceeds the natural flux 80-fold, and the ratio of man-made to natural flux is also large for a number of other cases (Table 10). This explains, in part, why toxic metal pollution was cited by 35 of 41 states that reported water quality problems to the Environmental Protection Agency in 1976 (Council on Environmental Quality, 1978).

Similarly, persistent pesticides consisting of chlorinated hydrocarbons, though banned for some time because of potentially harmful ecosystem impacts, are found with a 68 per cent detection rate in water and sediment samples in Houston, Texas. (Council on Environmental Quality, 1978). And DDT, while controlled in the U.S.,

TABLE 10

*Average Ratios Between Man-made and Natural Flux of Selected
Heavy Metals in the Global Environment*

Element	Ratio of man-made to natural flux
Nickel	0.9
Vanadium	1.3
Copper	2.3
Arsenic	3.3
Tin	3.5
Zinc	4.6
Cadmium	5.2
Selenium	14
Antimony	28
Molybdenum	29
Lead	70
Mercury	80

Source: Modified from Stumm (1977).

is increasingly being produced for sale in developing countries
(Goldberg, 1976).

Finally, acid rain – containing sulphuric acid resulting from fossil
fuel combustion in urban centres – is having a number of effects.
One of the most remarkable and potentially hazardous of these is
apparently a complete shift in forest-floor mineral cycling processes,

TABLE 11

Technology and Biology: A Checklist of Concerns

---

1. Habitat Modification

  - Reduction in biological diversity: e.g. replacement of natural species by hybrids that are less resistant to environmental variation.
  - Unexpected biological succession: e.g. river impoundments that produce explosive growth of exotic organisms, creating new habitats that may have harmful effects.
  - Alteration of regional and global bio-geochemical cycles: e.g. carbon dioxide, nitrogen, heavy metals, etc.

2. Genetic Effects

  - Development of resistance by potentially harmful organisms: e.g. resistance of bacteria to antibiotics, resistance of insects to pesticides.
  - Spread of potentially harmful mutations: e.g. through experimentation with recombinant DNA.
  - Reproductive failure in humans or other organisms: e.g. through exposure to sublethal amounts of certain synthetic organic compounds.

3. Spread of Disease

  - Rapid world-wide spread of disease organisms: e.g. spread of new strain viruses through travel and trade by air, land and sea.
    Increases in diseases related to urban conditions: e.g. gonorrhea and lung-cancer.
  - Reduction of disease defence mechanisms: e.g. through reduction in natural defence mechanisms, or through specific immunization with live vaccines.
  - Emergence of new diseases: e.g. as in the case of Legionnaires Disease or the increased evidence of cancer in aquatic organisms inhabiting polluted habitats.

---

which may eventually lead to problems with nutrient availability and metal toxicity, as well as direct damage to leaf tissue. (Seliga and Dochinger, 1976; Cronan *et al.*, 1978).

Thus, for ecosystems as for human mortality, we observe a range from acute to chronic effects, from easily understood to complex causal structure. Much of what is happening in ecosystems is in fact so incompletely understood that no clear cut directives can flow from scientific work to hazard management. All that science can presently hope to provide are warnings of what is possible. To illustrate this point a listing of some potential hazards for the future is given in Table 11.

Summary and Conclusions

Hazards arising explicitly or implicitly out of technological practices have in the industrialized world significantly surpassed natural hazards in impact, cost and general importance. At present in the United States technological hazards account by our estimate for 15-25 per cent of human mortality, with associated economic costs and losses of $50-75 thousand million annually. About half of these costs and losses are in connection with accidents and violence, the remainder with various forms of chronic disease. Ecosystem impacts, though difficult to define fully, are indicated by a number of danger signals, such as significant exogenous extinction of species, productivity losses and high concentrations of anthropogenic toxic chemicals in the environment. The impacts of technological hazards are clearly incomplete; serious consequences not considered in this paper, arise from threats to economy, society and environment.

Overall, the burden of risk assessment, hazard management, coping and adjustment may be as high as $200-300 thousand million per year, or 10-15 per cent of the Gross National Product. So far, the principal result of this effort has been the elimination of numerous acute effects such as infectious disease and point-source pollution, with little progress in stemming the tides of chronic disease and ecosystem impacts.

We conclude, therefore, that while the problem of technological hazards is on balance not getting worse, the main success of hazard management has so far been with the relatively more accessible part of the problem. And while this part of the problem is by no means under control, as seen by the continuing burden of violence and accidents, the principal challenge for the future involves hazards that have indistinct cause and a broad distribution of impacts. Coping with technological hazard is and will continue to be one of the major social issues of our time.

Acknowledgements

We thank Robert Goble, Thomas Hollocher, Jeanne Kasperson, Roger Kasperson, and Jim Tuller, all colleagues in the Clark Hazard Assessment Group, for providing interesting insights and specific information in the course of writing this article. We are also most grateful to Dennis Chinoy, Thomas Hollocher, Henry Kissman, Leslie Lipworth, David Pimentel, and Christopher Whipple who served as reviewers and provided helpful commentary. The research of the group was supported in part by the National Science Foundation under grant number ENV 77-15334. Any opinions, findings, conclusions or recommendations expressed herein are those of the authors and do not necessarily reflect the view of the National Science Foundation.

References

Bolin, B. (1977).   Changes of land biota and their importance for the carbon cycle. *Science* **196,** 613-615.

Brunnel, F., and Brunnel, P. (1967). "Extinct and Vanishing Animals." Springer Verlag, New York.

Burton, I., Kates, R.W. and White, G.F. (1978).   "The Environment as Hazard." Oxford University Press, New York.

Council on Environmental Quality (1975).   Environmental Quality 1974. U.S. Government Printing Office, Washington, D.C.

Council on Environmental Quality (1977).   Environmental Quality 1976. U.S. Government Printing Office, Washington, D.C.

Council on Environmental Quality (1978).   Environmental Quality 1977. U.S. Government Printing Office, Washington, D.C.

Cronan, C.S., Reiners, W.A., Reynolds, R.C. and Lang, G.E. (1978). Forest floor leaching: contributions from mineral, organic and carbonic acids in New Hampshire sub-alpine forests. *Science* **200,** 309-311.

Ehrlich, P.R., Ehrlich, A.H. and Holdren, J.P. (1977).   "Ecoscience: Population, Resources and Environment."   W.H. Freeman and Co., San Francisco, California.

Environmental Protection Agency. (1977a).   National Water Quality Inventory – 1976 Report to Congress. (EPA-440/9-76-024). U.S. Government Printing Office, Washington, D.C.

Environmental Protection Agency. (1977b).   National Air Quality and Emission Trends Report – 1976.   (EPA-450/1-77-002). U.S. Government Printing Office, Washington, D.C.

Epstein, S.S. (1976).   The political and economic basis of cancer. *Technology Review*, July-August, 35-43.

Faigin, B.M. (1976).   Societal Costs of Motor Vehicle Accidents. (Report DOT-HS 802 119).   U.S. Department of Transportation, Washington, D.C.

Fiering, M.B. and Holling, C.S. (1974).   Management Standards for Perturbed Ecosystems. *Agro-Ecosystems* **1,** 301-321.

Fisher, J., Simon, N. and Vincent, J. (1969). "Wildlife in Danger." Viking Penguin, New York.

Fraumeni, J.F., Jr. (ed.). (1975). "Persons at High Risk of Cancer: An Approach to Cancer Etiology and Control." Academic Press, New York.

Goble, R.L., Yersel, M. and Hohenemser, C. (1978).   "Estimates Direct Costs and Productivity Losses Due to Technological Hazard Mortality" (working paper).   Hazard Assessment Group, Clark University, Worcester, Massachusetts.

Goldberg, E.D. (1976).   "The Health of the Oceans."   UNESCO Press, Paris.

Higginson, J. (1976).   A hazardous society?   Individual versus community responsibility in cancer prevention. *American Journal of Public Health* **66,** 359-366.

Kates, R.W. (1978). "Comparative Hazard Sources in U.S. and Developing Countries" (working paper). Hazard Assessment Group, Clark University, Worcester, Massachusetts.

Likens, G.E., Bormann, F.H., Pierce, R.S. and Reiners, W.A. (1978). Recovery of a deforested ecosystem. *Science* **199,** 492-496.

National Institute of Health (1970). U.S. Cancer Mortality by County, 1950-1969. U.S. Government Printing Office, Washington, D.C.

National Safety Council. (1977). Accident Facts. National Safety Council, Chicago, Illinois.

Odum, H.T. and Odum, C.E. (1976). "Energy Basis for Man and Nature." McGraw Hill, New York.

Porter, P.W. (1976). "Agricultural Development and Agricultural Vermin in Tanzania" (conference paper). Department of Geography, University of Minnesota, Minneapolis, Minnesota.

Preston, S.H. (1976). "Mortality Patterns in National Populations." Academic Press, New York.

Rice, D.J., Feldman, J.J. and White, K.L. (1976). "The Current Burden of Illness in the United States" (occasional paper). Institute of Medicine, National Academy of Sciences, Washington, D.C.

Seliga, T.A. and Dochinger, L.S. (eds.). (1976). "Proceedings of the First International Symposium on Acid Precipitation and the Forest Ecosystem." U.S. Department of Agriculture, Washington, D.C.

Smithsonian Institution. (1975). "Report on Endangered and Threatened Plant Species in the United States". U.S. Government Printing Office, Washington, D.C.

Sommer, A. (1976). Attempt at an assessment of the world's tropical moist forests. *Unasylva* **28,** 5-25.

Spiegelmann and Erhardt (1974). "Mortality and Morbidity in the United States." Harvard University Press, Cambridge, Massachusetts.

Stumm, W. (ed.). (1977). "Global Chemical Cycles and their Alterations by Man." Abakon Verlagsgesellschaft, Berlin.

Tuller, J. (1978). "The Scope of Hazard Management Expenditure in the U.S." (working paper). Hazard Assessment Group, Clark University, Worcester, Massachusetts.

Vetz, G. and Johnson, D.L. (1974). Breaking the web. *Environment* **16,** 31-39.

Woodwell, G.M., Whittaker, R.H., Reiners, W.A., Likens, G.E., Delwiche, C.C. and Botkin, D.B. (1978). The biota and the world carbon budget. *Science* **199,** 141-146.

World Health Organization. (1976). "World Health Statistics Annual", Vol. I. World Health Organization, Geneva, Switzerland.

Woytinsky, W.S. and Woytinsky, E.S. (1953). "World Population and Production." Twentieth Century Fund, New York.

HEALTH AND ECONOMIC-DEVELOPMENT FACTORS AFFECTING MORTALITY

L.A. Sagan* and A.A. Afifi§

*Electric Power Research Institute,
Palo Alto, California, U.S.A.

§School of Public Health, U.C.L.A.,
Los Angeles, California, U.S.A.

The relationship between economic development and population size has been of interest ever since it became clear to Malthus in 18th century England that both appear to increase simultaneously. Demographic transition theory holds that the former precedes the latter and does so in a certain specified manner. Primitive agrarian societies are viewed as those with high death rates balanced by high birth rates, thus maintaining stable numbers. Increased economic activity is associated with an initial reduction in death rates, but not birth rates, leading to an increase in numbers. It is widely conjectured that improved nutrition, better sanitation, and the increased availability of medical care are the responsible agents. After some interval, birth rates fall as urbanization increases, women enter the labour force, and the economic advantages of large families disappear. The population then stabilizes at a level higher than that of the earlier agrarian phase.

This theory, while explaining much demographic history, is attacked on two grounds. First, it does not exactly describe all demographic experience. The second objection is that there is evidence that death rates can fall without a major reorganization of a peasant economy. In other words, substantial economic improvement may be a sufficient condition for a decline in mortality, but it is not today a necessary condition (Coale, 1975).

This study explores the relationship between industrialization and health. Our analysis is based on data from 150 countries, covering more than 99 per cent of the world's population.

Methodology

Measurement of Industrial Development

Per caput commercial energy consumption (ENERGY) has been

used in demographic studies as an indicator of modernization (Weller and Sly, 1969; Christian *et al.*, 1977). For our purposes, it has a number of advantages over the more commonly used *per caput* gross national product (GNP):

1) The use of GNP entails arbitrary adjustments to convert currencies to a common unit, as well as adjustments for inflationary tendencies. Since it is measured in constant physical units, energy consumption requires no such adjustments.

2) The use of national energy consumption, the majority of which in each country is consumed in industrial and commercial activities, avoids the assumption implied in the use of GNP that improved health can be purchased.

3) In an era of resource scarcity, when there are efforts afoot to "decouple" energy consumption from GNP, information about the relationship of health to energy consumption has its own inherent interest. The extent to which energy consumption can be restrained or reduced without affecting health is itself an important question.

4) There are other defects inherent in the use of GNP as a measure of economic activity: it fails to take account of work that occurs outside of the market (e.g. housework). It also fails to reflect improvement in quality and/or reductions in price that occur over time.

Hauser (1960) states that "The availability of non-human energy for the production of goods and services is perhaps the best single measurement available of differences in capital investment, know-how, and technology which account for the great differences in productivity and, consequently, in the size of the aggregate product available for distribution". Furthermore, increased commercial energy consumption is considered an indicator of the existence and effective utilization of more modern forms of the division of labour and other aspects of social organization (Cottrell, 1955; Gibbs and Martin, 1962). In this sense the effect of energy consumption on health can be viewed as largely indirect.

Measurement of Health

Life expectancy from birth is a statistic calculated from age-specific mortality rates at a given time point. A common misunderstanding of life expectancy is that children born at that point will experience those mortality rates throughout the subsequent years of their lives, an assumption likely to be in error. Life expectancy is, nevertheless, a very useful measure of mortality, since it is an aggregate of death rates currently being experienced. Infant mortality, or death within the first year of life, is also considered a useful measure of the level of health within a community. We have used both of these quantitites as measures of health in the following analyses. Figures 1

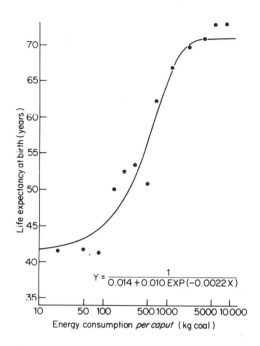

$$Y = \frac{1}{0.014 + 0.010\,\mathrm{EXP}(-0.0022\,X)}$$

Fig. 1 Relationship between energy consumption and life expectancy for 130 countries (1975 data).

and 2 illustrate the relationship between energy consumption (as kilograms of coal-equivalent, *per caput*, per year) and measures of health for all 130 countries for which complete data were available in 1975. Each data point shown is the average value for ten successive countries ranked in order of increasing ENERGY. The curves were fitted by an iterative procedure to a logistic function. Although the data could also be fitted to a linear function, a number of considerations made the logistic function preferable. As will be described later, there is no significant correlation between energy consumption and measures of health for either the countries with ENERGY below 100 kg coal or for the countries with ENERGY above 2000 kg coal, thus giving support to the use of a logistic equation. We interpret these relationships as demonstrating distinct phases in economic development as it affects mortality, namely from 0-99 kg coal (Phase 1), 100-1999 kg coal (Phase 2), 2000-3999 kg coal (Phase 3), and greater than 4000 kg coal (Phase 4). Life expectancy is relatively insensitive to increasing energy consumption throughout the first and fourth phases, but rises sharply through Phase 2 and levels off through Phase 3. A similar but inverse relationship holds for infant mortality. We further divide Phase 2 into two categories so as to examine the "take-off" phase of economic development in

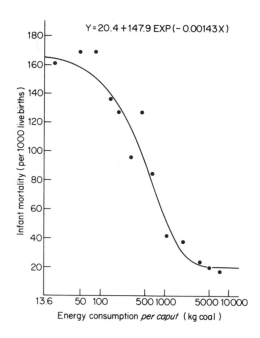

Fig. 2 Relationship between energy consumption and infant mortality for 130 countries (1975 data).

more detail: Phase 2a from 100-399 kg coal and 2b from 400 to 1999 kg coal.

The 29 countries in Phase 1, containing 9 per cent of the world's population, are characterized by illiteracy, inadequate and monotonous diet, and high rates of infant mortality and fertility. The primary economic activity is subsistence farming.   There is little exchange of money and energy inputs are largely from human muscle, i.e. non-commercial.   Ethiopia, Nepal and much of Africa below the Sahara are examples.   Phase 2 contains 63 per cent of the world's population.   *Per caput* GNP has begun to rise.   The portion óf the labour force engaged in agriculture begins to fall.   Urbanization commences.   Infant mortality starts to fall and longevity increases by approximately 28 years.   India, China and much of Latin America are examples.   In Phase 3, all of these trends continue although only small gains in longevity occur.   In Phase 4, the development process is mature.   Literacy is now virtually universal and health benefits plateau in spite of progressive industrialization.

Other Variables

The relationships in Figures 1 and 2 should not be construed as indicating causation.   Rather, it was our intent to examine many variables for which data are available for most countries in an attempt to clarify the relationship between economic development and health. We have carried out both "cross-sectional" and "longitudinal" studies as outlined below on 150 countries, which cover more than 99 per cent of the world's population.

The variables examined, the shorthand notations used in text and tables and their sources are listed below:

INFMORT	= infant mortality per 1000 live births (U.N. Demographic Yearbook).
LONGBI (e_o)	= life expectancy at birth in year (U.N. Demographic Yearbook).
FERTILITY	= annual births per woman aged 15 to 45 (U.N. Demographic Yearbook.
BIRTH RT	= live births per thousand population (U.N. Demographic Yearbook).
DEATH RT	= deaths per thousand population (World Bank, 1975).
GNP	= *per caput* gross national product in dollars (World Bank, 1975).
ENERGY	= *per caput* annual commercial energy consumption in kg coal equivalent (U.N., 1976).
% LABAG	= per cent of labour force employed in agriculture (Taylor and Hudson, 1972).
% URBAN (or % GTH)	= per cent of total population living in cities with greater than one hundred thousand, calculated from statistics (U.N. Statistical Yearbook).
% LITERACY	= per cent of persons over age 15 able to read and write (Banks, 1971; U.S. Agency for International Development, 1974).
ENROLM	= per cent of eligible population enrolled in school (Taylor and Hudson, 1972).
TOTCALS	= *per caput* total daily calories (FAO, 1971).
CARBCALS	= *per caput* daily calories from carbohydrate (FAO, 1971).
FATCALS	= *per caput* daily calories from fat (FAO, 1971).
PRTCALS	= *per caput* daily calories from animal and vegetable protein (FAO, 1971).
APRCALS	= *per caput* daily calories from animal protein (FAO, 1971).
POP/MD	= number of persons per physician (U.N. Statistical Yearbook).
POP/BED	= number of persons per hospital bed (U.N. Statistical Yearbook).
GINI	= an index of distribution of income (Taylor

TABLE 1

Means of Selected Variables By Phase of Development
(1975 data)

Variables	1 (N=29)	2a (N=34)	2b (N=41)	3 (N=16)	4 (N=21)	Total (N=150)
Total population*	365	1113	1337	396	798	3954
Death rate	23.0	15.8	10.5	8.5	9.4	12.76
Birth rate	47.4	41.5	30.9	20.0	16.5	31.7
Fertility	222	202	176	85	77	148
INFMORT	158.6	128.6	67.0	21.0	20.8	80.0
LONGBI (e_o)	41.3	50.6	60.9	71.0	71.0	59.0
GNP	120	160	459	2615	3967	1219
ENERGY	51.0	201	700	3277	7209	2022
% LABAG	72.3	68.2	63.7	28.3	21.0	52.6
% URBAN	6.7	13.3	30.3	41.5	45.3	26.4
% Literacy	21.2	36.0	68.2	93.7	98.4	62.9

Energy Phase

TABLE 1 (Contd)

TOTCALS	2092	1979	2201	2673	3131	2354
CARBCALS	1686	1545	1665	1691	1668	1636
FATCALS	295	257	350	619	1043	480
PRTCALS	220	200	243	314	364	259
APRCALS	36	31	77	123	202	93
POP/MD	28234	9355	2812	745	564	7067
POP/BED	2838	1496	540	121	106	1004

*In millions

	and Hudson, 1972).
% HALF	= smallest per cent of population receiving half of total income (Taylor and Hudson, 1972).
% CHRISTIAN	= per cent of population who are Christian (Taylor and Hudson, 1972).
% MOSLEM	= per cent of population who are Moslem (Taylor and Hudson, 1972).

Data Analysis

We obtained 1975 values of the above mentioned variables, whenever possible, for 150 countries. While information on population and vital statistics was available for each country, other data were incomplete. For instance, GINI and % HALF were available for only 71 countries; however, 116 countries had complete data on the variables used in most of the data analysis.

Results based on these data constitute our cross-sectional analysis. Statistical averages (weighted by population) were obtained using complete data on each variable. These averages are presented in Table 1 by energy phase. Standardized scores, i.e. (phase mean - grand mean)/total standard deviation, are shown in Figure 3. The signs of these scores were adjusted so that they were all positive for the developed phases. Figure 3 shows essentially horizontal bands, each corresponding to an energy phase. This indicates the strong interdependence among these variables, and hence the difficulty of our task.

Weighted simple correlations were estimated from complete pairs of observations. Weighted multiple regression analysis was based on the 116 countries with complete data. This latter analysis excluded % LABAG, ENROLM, GINI and % HALF. The longitudinal analysis described below is based on a subsample of 42 countries for which information was available in 1950 and 1970.

In the regression and correlation analysis some variables were transformed to their logarithmic values since this procedure improved linear correlation with the health measures. Variables so transformed were GNP, ENERGY, POP/MD and POP/BED. Finally, to further illustrate the degree of linear dependence between the health measures and the explanatory variables, we chose to report the proportion of standard deviation (SD) explained, rather than the familiar proportion of variance, since the SD is in the same units as the original measurements. While the proportion of variance explained is the square of the multiple correlation coefficient (R^2, the proportion of SD explained is $1 - 1-R^2$. The latter is always less than or equal to the former. For example, if $R = 0.8$, then the proportion of variance explained is 0.64 (64%) while the proportion of SD explained is 0.40 (40%).

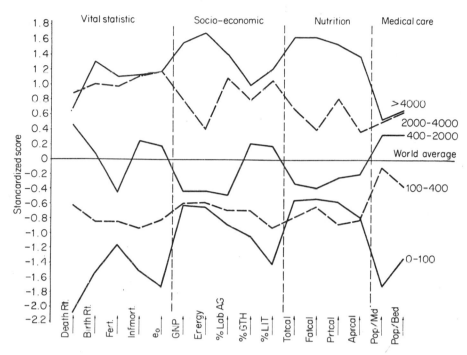

Fig. 3 Standardized scores for 16 variables, by annual *per caput* energy consumption (kg coal equivalent), for 150 countries (1975 data).

Results

Results will be presented in two sections, cross-sectional and longitudinal.

Cross-sectional Analysis

Correlation coefficients for selected variables with life expectancy and infant mortality are shown in Tables 2 and 3 for the total sample as well as within energy phases. We discuss these in groups of variables. The discussion focuses on life expectancy but similar interpretations are valid for infant mortality.

Vital Statistics Correlations between life expectancy and infant mortality are very high since death rates at all ages tend to vary simultaneously. Variations in infant mortality are therefore reflected in similar variations of longevity in the opposite direction. Death rates per thousand are inversely related to life expectancy in the poorer countries where the populations are relatively young

TABLE 2

Correlation Coefficients: Selected Variables with Life Expectancy
By Phase of Development
(1975 data)

Variables	Energy Phase					Total
	1 (N=29)	2a (N=34)	2b (N=41)	3 (N=16)	4 (N=21)	(N=150)
Vital Statistics:						
Birth rate	-.72	-.31	-.70	-.79	(-.25)*	-.92
Fertility	-	(.07)	-.80	-.69	(.19)	-.92
Death rate	-.99	-.97	-.93	(-.29)	.55	-.91
INFMORT	(-.10)	-.86	-.85	-.92	-.79	-.96
% under 15	-	.26	-.67	-.62	-.44	-.88
Nutrition:						
TOTCALS	(.10)	.31	(.17)	(-.04)	(-.16)	.73
% CARB	(.01)	-.38	(-.19)	(-.31)	-.62	-.65
FATCALS	(-.04)	.37	.30	(.16)	.65	.68
PROTCALS	(-.34)	(.15)	.33	(.06)	(-.12)	.75
APRCALS	(-.01)	.42	.57	(.02)	.47	.81

TABLE 2 (Contd)

Education:						
% Literacy	.54	.73	.76	.46	-.53	.95
ENROLM	-.46	.30	.62	(-.39)	.31	.63
Medical Care:						
POP/MD	(-.04)	-.38	-.82	(-.12)	.72	-.88
POP/BED	-.59	-.36	(-.38)	-.57	(.02)	-.91
Economic:						
GNP	(.20)	.29	(.04)	.87	.50	.85
Energy	(.32)	.52	.30	-.37	(.07)	.92
Social:						
GINI	--	.88	(-.04)	-.92	(.20)	-.39
% HALF	--	.85	(-.05)	-.93	(-.27)	.33
% Christian	(.34)	.37	(.19)	(-.14)	.59	.51
% Moslem	-.71	(-.23)	-.54	(-.22)	-.57	-.54
Urbanization:						
% Urban	(.23)	(.09)	.45	(.45)	(.22)	.78

*Coefficients shown in parentheses are not significant at the 0.05 significance level.

TABLE 3

Correlation Coefficients: Selected Variables With Infant Mortality, By Phase of Development
(1975 data)

Variables	Energy Phase					Total
	1 (N=29)	2a (N=34)	2b (N=41)	3 (N=16)	4 (N=21)	(N=150)
Vital Statistics:						
BIRTH RATE	(.36)*	(.11)	.83	(.40)	.39	.92
FERTILITY	(.43)	(-.03)	.68	.55	(-.09)	.91
DEATH RATE	(.05)	.87	.74	.56	-.53	.86
LONGBI	(-.10)	-.86	-.85	-.92	-.79	-.96
% UNDER 15	(-.10)	-.43	.50	(.48)	(.28)	.86
Nutrition:						
TOTCALS	.49	-.27	.24	(.43)	.16	-.67
% CARB	(.15)	(.05)	-.33	(-.21)	.89	.58
FATCALS	.52	-.27	(.16)	(.33)	-.89	-.63
PROTCALS	.55	(.02)	(.08)	(.23)	(.11)	-.70
APRCALS	(-.21)	-.48	-.38	(.07)	-.79	-.82

TABLE 3 (Contd)

Education:						
% LITERACY	-.55	-.80	-.83	(-.45)	.33	-.96
ENROLM	-.70	-.49	-.65	(-.31)	-.58	-.75
Medical Care:						
POP/MD	.67	(.16)	.63	(-.17)	-.79	.89
POP/BED	(-.20)	.42	(.14)	(.47)	(-.23)	.86
Socio-economic:						
GNP	.47	-.42	-.49	-.71	-.83	-.80
ENERGY	(.17)	-.52	(.02)	-.23	-.50	.88
GINI	-	-.95	(.03)	.90	(-.16)	.36
% HALF	-	.93	(.07)	-.93	(.20)	-.31
% CHRISTIAN	(.16)	-.48	(.21)	.60	-.75	-.44
% MOSLEM	(.05)	(.00)	(.73)	(.33)	.79	.52
Urbanization:						
% GTH	(.27)	(-.06)	-.46	-.64	-.54	-.78

*Figures in parentheses are not statistically significant at the 0.005 level of significance.

in age distribution.   With increasing economic development, death rates fall at younger ages, the mean population age increase and life expectancy becomes positively correlated with death rates in these older populations.   As predicted by demographic transition theory, poorer countries with high death rates have high birth rates and high fertility in order to maintain their population size, thus explaining the negative correlation of birth rate with life expectancy.   In Phase 4 countries, this correlation becomes non-significant as birth rates drop to replacement levels.

Nutrition Of the five nutritional variables tested, all were highly correlated with life expectancy and infant mortality in the overall sample;   however, absolute correlations within energy groups were either low or non-significant.   The carbohydrate variable is negatively correlated with life expectancy since it is calculated as the percentage of total calories from carbohydrate, rather than measured in absolute levels as with the other nutritional variables.

Education Several measures of educational levels among countries were examined but excluded from further analysis because of poor statistical correlation with health.   These included dollar investment in education *per caput* and numbers of scientific publications.   The most powerful explanatory variable was literacy, the ability to read and write.   The overall correlation coefficient between literacy and our two measures of health were 0.95 and -0.96, respectively.   Furthermore, correlations were highly significant within each of the energy groupings with the exception of Phase 4 where, with literacy approaching 100 per cent, literacy no longer serves as a useful discriminatory variable.   Correlations for the enrollment variable were weaker and less consistent within energy phases.

Medical Care The effect of medical care was measured by two variables, the ratios between the total population and the number of physicians and of hospital beds, respectively.   Both ratios fall rapidly with economic development as seen in Table 1.   Overall correlations are highly significant;   however, within energy phases, correlations are neither strong nor consistently significant.   In Phase 4 the correlation, 0.72, becomes positive for the ratio of population to physicians, i.e. the smaller the number of physicians, the higher the life expectancy within those 21 countries.   Number of hospital beds does not correlate significantly with health in this same group.

Economic Energy consumption *per caput* and GNP *per caput* are both measures of economic development, and both correlate highly with health.   Of the two, correlation with energy consumption is higher in Groups 2a and 2b where the greatest increase in life expectancy occurs, whereas GNP becomes more strongly correlated in Groups 3 and 4 where increments to expectancy are small.

Social Variables Two measures of the distribution of wealth within countries were examined to explore whether an equitable distribution might not have a beneficial effect upon health. Neither variable, GINI index nor % HALF, confirmed this expectation, both showing weak correlations.

Urbanization We examined a number of measures of urbanization, using various criteria of urban conglomeration, including the index of urbanization provided to the United Nations by each individual government. Per cent of population in cities greater than 100,000 was the measure of urbanization most highly correlated with health. This variable increases from an average of 6.7 per cent in Phase 1 to 45.3 per cent in Phase 4 (Table 1). Although the magnitude of the overall correlation is .78, for both life expectancy and infant mortality, correlations within energy phases are either not significant or only slightly so.

Other Variables Religion, through its injunctions on food, sanitary practices and values, might be expected to have significant effects on health. Overall correlations for per cent Christian and per cent Moslem were moderately strong, but because no significant correlations appeared within groups, it was concluded that these variables offered little as explanatory variables and they were dropped from further analyses.

Multiple Regression

In order to evaluate the relative contribution of our independent variables to health, we grouped them into six sets, selecting those variables which had demonstrated highest correlations. They are: Nutrition (five variables as in Tables 2 and 3); medical (POP/MD, POP/BED); literacy; energy consumption; degree of urbanization (% GTH); and literacy and energy combined. Other variables showing weak correlations were excluded. Multiple regressions were carried out between life expectancy (e_o) and each set of variables, as well as partial correlations with the remaining variables. For each set of variables, the per cent of SD of e_o explained is shown in Table 4. After removing the linear effect of a given set of variables, variables with partial correlation greater than 0.5 in magnitude are also indicated in Table 4. Table 5 shows the results of a similar analysis with infant mortality used as the dependent variable. This analysis was performed for the total sample as well as for each energy phase with the exception of Phase 1 where data were insufficient.

TABLE 4

Variation of Life Expectancy Among Nations
Explained by Selected Variables
(1975 data)

Variables	Energy Phase				
	2a (SD=3.7)	2b (SD=7.0)	3 (SD=2.2)	4 (SD=1.2)	Total (SD=11.4)
Nutritional: (NUTS)	10% LIT*	22% MD,LIT	0% BEDS LIT GNP	35%	53% MD,BEDS LIT,GNP EN
Medical: (MD, BEDS)	12% LIT	41%	10% GNP,EN	37%	67%
Literacy: (LIT)	30% NUTS	29% MD,GNP	3% BEDS	14% MD,NUTS	67% GNP,EN
Energy Consumption: (EN)	17% LIT	16% NUTS,MD LIT	0% BEDS BNP	0% MD,LIT NUTS	65% LIT

TABLE 4 (Contd)

Urban: (GTH)	0%	8%	5%	0%	37%
	LIT,EN	NUTS,MD	BEDS	MD,NUTS	NUTS,MD
		LIT	GNP	GNP	BEDS,LIT
					GNP,EN

Literacy and Energy:	35%	34%	3%	13%	73%
		MD,GNP	GNP	MD,NUTS	
			BEDS		

*After removing the linear effect of the nutritional variables, % literacy has a partial correlation greater than 0.5 in absolute values. Other entries are to be similarly interpreted.

TABLE 5

Variation of Infant Mortality Among Nations
Explained by Selected Variables
(1975 data)

Variables	Energy Phase					
	1 (SD=23.2)	2a (SD=24.2)	2b (SD=32.7)	3 (SD=11.3)	4 (SD=4.6)	Total (SD=57.1)
Nutritional: (NUTS)	5% LIT* MD	12% LIT	20% LIT	3%	54%	52% LIT MD EN
Medical: (MD)	22%	10% LIT	20% LIT	5%	37% NUTS EN GTH	60% LIT
Literacy: (LIT)	11% NUTS GNP MD	40% NUTS	62%	3% GNP GTH	4% MD NUTS	72%

TABLE 5 (Contd)

Energy: (EN)	0%	16%	13%	0%	11%	61%
	NUTS	LIT	LIT	GNP	MD	LIT
	MD			GTH	NUTS	
					GNP	

Urban %: (GTH)	0%	0%	9%	19%	15%	37%
	NUTS	LIT	LIT		MD	LIT
	MD	EN			NUTS	EN
	LIT				GNP	MD
						NUTS

*After removing the linear effect of the nutritional variables, % literacy has a partial correlation greater than 0.5 in absolute values. Other entries are to be similarly interpreted.

For the total sample, three of the grouped variables are able to explain approximately two-thirds or more of standard deviation in longevity: medical, literacy and energy consumption. Literacy and energy each retain significance after removal of the other but the two together explain only slightly more of the standard deviation of life expectancy than does either alone.

Medical variables are not only very powerful in reducing standard deviation in the overall sample but also are consistent within energy phases. In the 2b group of nations where life expectancy is rapidly increasing, medical care explains 41 per cent of standard deviation, and no other set of independent variables has a large partial correlation after the medical care effect is removed.

Urbanization is a very weak variable in this analysis, both in the total sample and in individual energy phases, where it is almost useless in explaining variability.

In summary, all of the variables examined demonstrated some relationship with health. With the exception of literacy, energy and medical care, the relationships were neither strong nor consistent enough to suggest a causal relationship. We therefore undertook further studies in order to determine whether these variables have predictive power.

Longitudinal Analysis

We collected data, where available, for the years 1950 and 1970. The sample size, 42, is considerably smaller than that for the cross-sectional analysis because of absence of data for earlier decades. Statistical means of selected variables are shown in Table 6. Total calorie consumption increased by 7 per cent, accompanied by a decrease in carbohydrate and an increase in fat calories. Total dietary protein remained relatively constant but daily calories from animal sources increased by 22 per cent. There were increases in energy consumption, literacy, availability of medical care and urbanization.

Correlations with life expectancy and infant mortality for each of these variables are presented in Tables 7 and 8. Correlations are high in both 1950 and 1970 with the exception of urbanization. These tables also show correlations of differences in the health measures, with differences in each of the variables, 1950-1970. The nutritional and medical variables were not significantly correlated with health. However, increases in life expectancy and decreases in infant mortality were significantly correlated with increases in both energy and literacy.

Discussion and Summary

As shown in Figure 3, all of the variables that we studied are closely associated with the process of economic development. In such a tight web of interrelationships, any one of those variables

TABLE 6

Statistical Means of Selected Variables
(42 countries)

	1950	1970	% Change
Longevity	51.3	63.1	+ 23
Infant mortality	111.6	57.8	- 48
% Literacy	62.3	76.4	+ 23
Energy	1276	2648	+ 108
TOTCALS	2706	2890	+ 7
FATCALS	333	980	+ 197
APRCALS	144	176	+ 22
CARBCALS	2058	1575	- 23
PROTCALS	315	326	+ 3
POP/MD	1751	1225	- 30
POP/BED	358	226	- 37
% Urbanized	25	36	+ 44

could be chosen as an index of development. Life expectancy itself served as an index of social progress as early as 1944 (Hart and Hertz, 1944).

We undertook an analysis of those relationships partly in order to test whether energy consumption was related to life expectancy closely enough to serve as a proxy or index for economic development. In the absence of experimental evidence, we believe that the results of our statistical analysis, both cross-sectional and longitudinal, are consistent with this hypothesis.

We have presented the relationship between commercial energy consumption and health as a logistic function. This data could have been fitted to a linear model as well but there are two reasons for rejecting a linear model:

1) There is no significant correlation between energy consumption and life expectancy or infant mortality below 100 or above 4000 kg coal equivalent. The improvement in health that has been observed longitudinally in the developed countries over the past decades can be explained through another mechanism discussed in a separate publication (Afifi and Sagan, in press).

TABLE 7

*Correlation Coefficients: Selected Variables
With Life Expectancy*

(1950-1970 data)

Variables	Life Expectancy (1950)	Life Expectancy (1970)	Differences (1950-1970)
% Literacy*	.98	.96	.77
Energy §	.90	.94	.45
FATCALS	.71	.75	(-.32)
PRTCALS	.70	.65	(.30)
APRCALS	.66	.75	(-.24)
TOTCALS	.76	.73	(.31)
POP/BED	-.89	-.92	(.24)
POP/MD	-.81	-.71	(-.06)
% GTH	.51	(.33)	(.08)

*Based on 42 countries, otherwise 22.
§Figures in parentheses are not statistically significant at the 0.5 level.

2) The linear model fails to pass the test of plausibility. Clearly, longevity cannot reach to infinity, but must have both lower and upper boundaries, conditions that are fulfilled by a logistic function.

Another purpose of this study was to examine those intermediate variables thought to underlie the relationship between increased life expectancy and economic development. The variable most significantly and consistently related to health in our analysis is literacy. Furthermore, although literacy clearly must be related to formal education they are not the same. Literacy was more highly consistent in its relationship to health than was school enrollment, although there may be a lag in the effect of school enrollment that partially masks its effect.

How should we interpret the effect of literacy? Is it simply an association without any causal significance? We think not: the correlations are too strong and too consistent. Although

TABLE 8

Correlation Coefficients: Selected Variables
with Infant Mortality

(1950-1970 data)

Variables	Infant Mortality (1950)	Infant Mortality (1970)	Difference (1950-1970)
% Literacy*	-.96	-.98	+.86
Energy §	-.89	-.86	+.83
FATCALS	-.76	-.78	(-.29)
PRTCALS	-.66	-.70	(+.39)
APRCALS	-.78	-.78	+.53
TOTCALS	-.83	-.77	+.69
POP/BED	+.80	+.92	(.06)
POP/MD	+.72	+.75	(-.33)
% GTH	-.57	-.37	(-.16)

*Based on 42 countries, otherwise 22.
§Figures in parentheses are not statistically significant at the 0.5 level.

there are undoubtedly many subtle interactions between health, economic development and education, we speculate that the effects on health which flow from education may fall into two categories: access to information and a change in values. The former includes access to nutritional, medical and other health materials and the latter, the abandoning of the detrimental practices associated with traditional cultures.

There is a second sense in which the significance of literacy may be interpreted and that is as a marker for that constellation of values and attitudes to which social scientists refer as "modernism." By modernism is meant rationalism, active participation in civic and national affairs, social, physical and intellectual mobility, and freedom from traditional mores (Inkeles and Smith, 1974). In analysing six Middle Eastern cultures, Lerner (1963) finds that the variable most highly correlated with these qualities is literacy, and that "Literacy is the basic personal skill that underlies the whole modernizing sequence."

Whereas the relation of maternal nutrition to pregnancy outcome remains unsettled (Thomson and Hytten, 1973; National Academy of Sciences, 1970) there can be little question that malnutrition plays a crucial role in infant mortality, particularly in developing countries. Often, the role of malnutrition is concealed in vital statistics data by infections such as diarrhoeas and pneumonias, which are far more common and severe in the malnourished child. Similarly, the resultant infection is likely to aggravate the level of malnutrition (Scrimshaw et al., 1968). Although dietary protein requirements are given careful attention by paediatricians, our data shows total calories to be more significant than protein calories as an indication of nutritional adequacy in national food supplies as measured by infant mortality rates. Latham (1975) points out that there has been an overemphasis on the importance of dietary protein, since, when the staple food is cereal, protein deficiency rarely occurs unless there is also a calorie deficiency.

Our analyses of nutritional data show high correlations with reductions in infant mortality. Both total calories and protein calories derived from animal protein maintain their strong correlation in longitudinal as well as in cross-sectional data. Yet we are reluctant to attribute much of infant mortality decline to this factor since we have so little evidence that infant nutrition relates directly to aggregated national data. On the other hand, examination of national nutritional data can be justified for the following reason. Although there is very little survey data on distribution of nutritional factors by age, or by economic status within nations, the data that do exist suggests that if there is not enough food for the whole family, the working adults tend to take for themselves the largest share (United Nations, 1974).

Critical to an evaluation of infant nutrition is an estimate of weaning age. The frequency of malnutrition and infectious disease and the consequent risk of death increase precipitously following weaning. Unfortunately, earlier weaning is apparently increasing throughout the world. Participation of mothers in the work force, urbanization, and imitation of more affluent classes all seem to underlie this phenomenon. In wealthier communities, families may substitute cow's milk and an otherwise adequate diet, whereas poorer families cannot afford to do so. For example, Reutlinger and Selowsky (1976) calculate that the unskilled working mother of a newborn infant in Calcutta must spend over 50 per cent of her income on cow's milk if she is to adequately replace the nutritional value of her own milk. We conclude that although the demonstrated relationships between national food supplies and infant mortality are strong, there are hidden intervening variables that are operating.

The availability of medical care, as reflected in our crude measure of population per doctor or per hospital bed, shows a very strong relationship between economic development and our health measures. Whereas this relationship is strong in the cross-sectional analysis overall, it is not significant in the longitudinal analysis or within phases in the cross-sectional analysis. We are reluctant to accept medical care as playing a strong role in the reduction of death rates for additional reasons:

1) Reduction in death rates began long before truly effective medical therapy became available. Death rates began falling in some areas of Europe at least one hundred years ago, and in most of Europe 50 years ago. Sulfa drugs were introduced in the 1930s and antibiotics in the late 1940s. Their use outside of the developed countries is even more recent. It seems unlikely in the extreme that the introduction of these drugs would have become so common throughout the underdeveloped world as to add many years to life expectancy within two decades. Furthermore, it would be an error to assume that these drugs are universally effective in treating and curing the diarrhoeas, pneumonias and other infections which, to a large extent, are responsible for death rates. Nor are these drugs easily used by unskilled persons. Since the population to physician ratio is of the order of many thousands in underdeveloped countries (Table 1), the burden of proof would seem to rest on those who allege medical care to be the cause of this phenomenon (McKeown et al., 1974).

2) Death rates have fallen at all ages proportionately. It would seem unlikely that medical care has had an effect equally on multiple diseases acting independently at different ages throughout life.

3) Since much of infant mortality beyond the immediate neonatal period is environmental in origin, therapeutic interventions are not likely to be effective. For example, 80 per cent of children between the ages of 7 months to 2 years admitted to a Johannesburg hospital were significantly malnourished (Wagstaff and Geefhuysen, 1974). A follow-up of children hospitalized in Iran revealed that one third had subsequently died (Sadre et al., 1973). Furthermore, few of those that survived had significantly improved their weight relative to their age after this period.

4) Examination of specific causes of death among infants reveals that in the neonatal period (first month) there are a variety of vague, ill-defined conditions, and congenital malformations. Only rarely are there specifically recognizable defects for which curative intervention is available. The very slow decline in infant mortality rates in developed countries is a reflection of this. Beyond the neonatal period, infections such as diarrhoeas and pneumonias are complications of other underlying pathology, frequently malnutrition. Studies have shown that the frequency of infectious diseases causing mortality among children show a significant decline when nutritional supplements are added

(Scrimshaw, 1966). When both medical care and nutritional supplements were provided in an experiement in India, the nutritional supplements had the major impact (Taylor and De Sweener, 1973).

On the other hand, several investigators have found significant effects of adequate obstetric and paediatric care in the developed countries where the major problem of nutrition has been overcome. For example, it is estimated that the infant mortality of 21.9 per thousand live births in New York City (1968) could have been reduced to 14.7 per thousand, a 33 per cent reduction, had all pregnant women received adequate care (National Academy of Sciences, 1973). Such studies, even when controlled for socio-economic and medical variables, demonstrate an effect of medical care in reducing infant mortality risk. In another New York study (Lee *et al.*, 1976) it was concluded that 75 per cent of the decline in infant mortality in the past decade was due to spontaneous reduction in the frequency of high risk, low birth weight children and that only 25 per cent of the improvement could be attributed to medical care. Thus, while medical care appears to be superceded in importance by nutrition in the developing countries, it may play some role in the more developed countries where nutrition is no longer a problem.

Another explanation of increased longevity is the effect of improved sanitation and preventive health. This seems plausible in view of the fact that reductions in death rates associated with economic development are often concurrent with reductions in the frequency of infectious disease. Variables which we tested for this effect: per cent of homes with piped water or flush toilets, showed no significant relationship to our health measures (data not shown). An extensive World Health Organization study shows improved sanitation to be a necessary but not sufficient condition for the reduction of death rates (Pineo and Subrahmanyam, 1975).

Urbanization appears to be one of the unalterable features of economic development. Although urban-rural mortality rates have often differed, there has been no consistent pattern. In Scandinavia, urban infant death rates were historically at least 50 per cent greater than rural infant death rates. That discrepancy gradually disappeared as urban death rates fell more rapidly than did rural rates, so that by 1960 they were identical (United Nations, 1973). In the developing countries today, infant mortality appears to be lower in cities than in rural areas. This has been shown in South America, Africa and Mexico (Johnson, 1964). Conflicting data on the effect of urbanization on infant mortality could well be the result of the simultaneous operation of a number of factors influencing infant mortality in different directions: crowding, employment of women, availability of medical care and differences in weaning practices.

In summary, economic development has been shown to add approximately 30 years to life expectancy and reduce infant mortality by 140 deaths per thousand live births. Literacy and energy consumption have demonstrated strong associations with health. We have suggested that literacy may operate through increasing access to information, or by association with other values, or both. *Per caput* commercial energy consumption was interpreted as a surrogate for industrialization.

References

Afifi, A.A. and Sagan, L.A. (1979). "Health and Economic Development: 2, A New Index of Health Development," (in press).

Banks, A.S. (1971). "Cross-polity Time Series Data." MIT Press, Cambridge, Mass. and London.

Christian, B., Ray, D., Benyoussef, A. and Tanahashi, T. (1977). Health and socio-economic development: an intersectorial model. *Social Science and Medicine* 11, 63-69.

Coale, A.J. (1975). The demographic transition. *In* "The Population Debate, Dimensions and Perspectives", Vol. 1. United Nations, New York.

Cottrell, F. (1955). "Energy and Society." McGraw-Hill, New York.

Food and Agriculture Organization. (1971). Agricultural Commodity Projections, 1970-1980, Vol. 2. (CCP 71/20). United Nations, Rome.

Gibbs, J. and Martin, W. (1962). Urbanization, technology and the division of labour: international patterns. *American Sociological Review* 27, 667-677.

Harte, H. and Hertz, H. (1944). Expectation of life as an index of social progress. *American Sociological Review* 9, 609-621.

Hauser, P.M. (1960). Demographic dimension of world politics. *Science* 131, 1641-1647.

Inkeles, A. and Smith, D.H. (1974). "Becoming Modern: Individual Change in Six Developing Countries." (PG 290) Harvard University Press, Cambridge, Massachusetts.

Johnson, G.Z. (1964). Health conditions in rural and urban areas of developing countries. *Population Studies* 17, 293, 309.

Latham, M.C. (1975). Nutrition and infection in national development. *In* "Food: Politics, Economics, Nutrition and Research" (P. Abelson, ed.). American Association for the Advancement of Science, Washington, D.C.

Lee, K., Tseng, P., Eidelman, A., Kandall, S. and Gartner, L. (1976). Determinants of the neonatal mortality. *American Journal of Diseases of Children* 130, 842-845.

Lerner, D. (1963). "Passing of Traditional Society: Modernizing the Middle East." Free Press, New York.

McKeown, R., Record, G. and Turner, R.D. (1974). An interpretation of the decline of mortality in England and Wales during the twentieth century. *Population Studies* **29,** 391-422.

National Academy of Sciences, National Research Council (1970). Maternal Nutrition and the Course of Pregnancy. National Academy of Sciences, Washington, D.C.

National Academy of Sciences, Institute of Medicine (1973). Infant Death: An Analysis by Maternal Risk and Health Care. National Academy of Sciences, Washington, D.C.

Pineo, C.S. and Subrahmanyam, D.V. (1975). Community Water Supply and the Excreted Disposal Situation in the Developing Countries. World Health Organization, Geneva.

Reutlinger, S. and Selowsky, M. (1976). Malnutrition and Poverty: Magnitude and Policy Options. (World Bank Staff Occasional Papers No. 23). John Hopkins Press, Baltimore and London.

Sadre, M., Donoso, G. and Hedayat, H. (1973). The fate of the hospitalized malnourished child in Iran. *Journal of Pediatrics Environment, Child Health* **19,** 28-31.

Scrimshaw, N.S. (1966). The effect of the interaction of nutrition and infection on the pre-school child. *In* Preschool Child Malnutrition, pp.663-74. National Academy of Sciences, National Research Council, Washington, D.C.

Scrimshaw, N.S., Taylor, C.E. and Gordon, J.E. (1968). Interactions of Nutrition and Infection. World Health Organization, Geneva.

Taylor, C. and Hudson, M. (1972). "World Handbook of Political and Social Indicators." Vale, New Haven, Connecticut.

Taylor, C. and De Sweemer, C. (1973). Nutrition and infection. *In* "Food, Nutrition, and Health" (M. Rechcigl, ed.), pp.204-225. S. Karger, Basel.

Thomson, A.M. and Hytten, F.G. (1973). Nutrition during pregnancy. *In* "Food, Nutrition and Health," (M. Rechcigl, ed.), pp22-45. S. Karger, Basel.

United Nations World Food Conference. (1974). Assessment of the World Food Situation, Present and Future. (Item 8 of the provisional agenda, paragraphs 160-161.). United Nations, Rome.

United Nations (1973). The Determinants and Consequences of Population Trends. pp.121-126. United Nations, New York.

United Nations (1976). World Energy Supplies, 1950-1974. (STAT/ (ST/ESA/STAT/SER. J/19). United Nations, New York.

United Nations (published annually). Demographic Yearbook. United Nations, New York.

United Nations (published annually). Statistical Yearbook. United Nations, New York.

United States Agency for International Development (1974). Annual Report of the Bureau for Population and Humanitarian Assistance, Fiscal Year 1973. U.S. Government Printing Office, Washington, D.C.

Wagstaff, L.A. and Geefhuysen, J. (1974). Incidence and spectrum of malnutrition in paediatric hospital wards.   *South African Medical Journal* **48,** 2595-2598.

Weller, R.H. and Sly, D.F. (1969).   Modernization and demographic change: a world view. *Rural Sociology* **34,** 313-326.

World Bank (1975).   World Bank Atlas 1975:   Population, *Per Capita* Product and Growth Rates. World Bank Group, Washington, D.C.

THE PROBLEM OF QUANTIFICATION

Torbjörn Thedéen

Department of Statistics, University of Stockholm, Stockholm, Sweden.

In order to quantify risk in energy production it is necessary first to define more closely the risk concept. It is evident that risk has to do with negative-valued events which are more or less uncertain or random. But if one considers the connection between risk and decision, or rather a sequence of decisions, it turns out to be of little use to put down just *one* definition of risk (Rowe, 1977). If we use for example the expected number of fatalities per terawatt hour (TWH) as a risk measure, it masks the fact that people react much more strongly to an identifiable catastrophic event than to a succession of fatalities more spread out in space and time and hence, less readily perceived. Therefore the decision situation, its elements and value hierarchy should indicate the risk concept to be used. This paper will point out the elements which are involved in risk: identification of hazards, risk takers, the negative consequences and lastly their likelihood.

In the discussion of risks involved in energy production there are two categories of primary interest:

1) possible adverse effects on health caused by chemical compounds in emissions to water and air – a problem of risk identification;

2) low probability catastrophes – a large number of fatalities and injuries concentrated in space and time.

Quantification is especially difficult in these two cases in contrast to such "conventional" risks as traffic- and occupational-accidents. We shall therefore deal more closely with the "unconventional" ones. Risk can be quantified by extrapolation of statistical data, by analysis of models and by more or less subjective judgements. All these estimates should be accompanied by some measures of their uncertainty, e.g. confidence intervals.

We shall also consider how risk estimates given by model

analysis can be compared with those given by statistical data, and here in particular look at a core melt-down in a nuclear reactor as an example.

Risk and Decision

Let us first point out that it is of no interest to study risk in isolation, as it is nearly always connected with decisions of some kind. This means that the definition of risk is apt to vary in relation to the corresponding decision and strategy. However, the concept of risk always includes uncertain or random events of negative value and some measure of their likelihood.

It is often natural to limit consideration to effects on human health and the environment, and furthermore the negative value of these consequences is usually regarded from the point of view of the decision-maker. It is often meaningful to consider the statistical probability of the events, but in cases such as sabotage and terrorism such a measure is of little use.

Let us first examine the decisions required. We start with some information about the activity in mind, e.g. energy production. We then formalize the situation, stating the alternatives and their consequences both deterministic and probabilistic in character and having associated with them positive and negative values. After such an analysis is performed a value structure is added and we then arrive at a decision, followed by a corresponding action. This action then creates a new decision situation at a later time and the above cycle of events is repeated for the new situation.

We thus seem to be moving spirally around and along a time axis towards the most satisfactory problem resolution.

Three groups of people are involved in a decision situation:

1) those getting the benefit of the activity;
2) those taking the risks;
3) the decision-making body

These groups can be more or less separate. It is however a typical feature of modern society that these groups are neither distinct nor easily identifiable, and this is certainly the case in the energy sector.

When we want to quantify risks in energy production it is not sufficient to be given only the number of fatalities etc. per TWH, since the risk-bearing groups might differ in substantial ways – as is the case with the workers inside and the population outside the power plants, as well as impacts on future generations.

To sum up, we need to quantify two aspects of risk. The first is risk identification. What constitutes a threat to health and life? And at which concentration? The second aspect is risk estimation. This includes the identification of risk-bearing groups and the estimation of the probability distribution of fatalities, injuries etc.

Identification and Estimation Methods

Let us first exclude the cases where risk is easy to identify – i.e. the many types of adverse health effects which turn up without any substantial time lag and are easy to diagnose. Again, for cases where we have long data records e.g. certain occupational diseases and accidents, no difficult quantification problems exist.

In other cases the situation is much more difficult. The problem with risk identification is mainly caused by the time lag. For example, carcinogenic and teratogenic effects, as well as potential changes of the climate caused by carbon dioxide from burning fossil fuels, take quite a long time to happen and to verify. There are nevertheless some ways to identify long-term health effects. The most "abstract" method is to make an expert judgement based purely on biochemical knowledge about the effect. The next step is to perform some experiments, probably on laboratory animals. Relatively large doses are often necessary in order to get any measurable effect in a short time. If we find any response we assume some plausible form of dose-response curve, probably linear (see Figure 1).

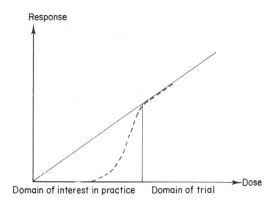

Fig. 1 Linear dose-response curve extrapolated from high-dosage experimental exposure (solid line), and probable dose-response curve under natural conditions (dotted line).

We then face two problems:

1) What is the actual form of the dose-response curve (for animals)?

2) Is it possible to extrapolate these findings to man?

To generalize rather broadly, it appears in most cases defensible to extrapolate to man from findings on animals. The form of the dose-response curve is still debatable, even in such a closely studied case as the carcinogenic effects of nuclear bombs. The extrapolation made seems to be more or less a matter of how careful one likes to be.

The other way to identify risks is to rely on non-experimental data about correlations between diseases and occupation or place of residence. Using data registers, information about certain diseases, in particular various forms of cancer, can be compared with data about where people work and live. The long census records in Sweden give us rather good material for a cancer-environment register, and one is now being constructed.

There is however a negative relationship between the usefulness of such data registers and the mobility of people in modern society. It is increasingly common that people change both occupation and residence many times during their lives. This makes it more difficult to discover potentially causal correlations between, for example, cancer incidence and a certain environment or profession. Furthermore, any effort to even out possible risks in work will make these risks even more difficult to identify and quantify.

Let us now turn to the estimation of risk, that is the probability of negative effects. It is important to remember that from a strictly scientific point of view risk is never known with certainty, even if we have a long data series. Estimation always includes assumptions about the future behaviour of the system being analysed – probability models, temporal consistency etc. Estimates are subject to two kinds of uncertainty: error due to wrong model assumptions, and statistical error due to fluctuations in the data base.

The second type is the one most discussed in statistical textbooks, and also the easiest to quantify, e.g. by means of confidence intervals containing the true value with a given high probability before the observations are made. The model-error is quite naturally more difficult to get hold of and to quantify. When estimating, it is therefore usually preferable to use methods which require fewer assumptions and more data, e.g. long data series and a simple assumption that the risk remains the same.

Three methods, sometimes in combination, are used in estimation:

1) Subjective or expert judgements;
2) Model analysis;
3) Estimates based on data series.

The typical risk estimation history for a production method or a technical system is a development from the first method to the second, and then the third. Consider as an example the nuclear reactor. When the first reactors started, safety estimations usually had to be based on expert judgements and some model calculations. When reactors came into commercial use and quite a few had been built, the risk of a possible catastrophe caused by a core melt-down became a widespread concern. The Rasmussen study (1975) was made, using a rather complicated model. It should however be noticed that as inputs to such a model one has to use elementary probability and parameter estimates obtained either from other applications or from expert judgements. The third phase in the history – estimation from data series – needs adequate data, and this is not yet available for nuclear reactors.

All estimates should be accompanied by some measure of their uncertainty. When the estimate is based mostly on judgements and model analysis, sensitivity analysis is often used, i.e. the effect of the other possible input values is analysed. If we use data series, uncertainty is estimated by confidence intervals.

One typical feature of modern society is that we have larger and more complex technical systems which often win widespread acceptance quite rapidly. This makes it more difficult to arrive at the third phase of the risk-estimation procedure described above. There are two ways to circumvent this difficulty and acquire more data rapidly:

1) register and analyse risk data on an international basis;
2) register and analyse incidents (e.g. break-downs of subsystems).

A good example of risk-estimation and risk-control along these lines is shown in the field of civil aviation.

An Example

It is important, but very difficult to estimate the probability of catastrophes for large, relatively new technical systems such as nuclear reactors. We here consider the probability of a core melt-down in a nuclear reactor, as the discussion may be valid for other technical systems as well. Before data has been accumulated by running reactors for some years it is of course necessary to rely on probability calculations as was done in the Rasmussen report. The components of the analysis in this report are:

1) Estimates of the probability of failure for each of the different components of the system. Such estimates are sometimes based on long data records from other uses of the components and in other cases on more or less empirical expert judgements.

2) A logical-model of the reactor can be used to point out the most likely pathways leading to a core melt-down.

The probability of a core melt-down was then estimated as a function of probability estimates for component failures. This resulted in an estimate for the core melt-down frequency (i.e. the probability/reactor year) of about 0.05 per cent, with an upper limit of 0.3 per cent.

We can also use empirical data to get an upper limit for the core melt-down frequency. Let t be the total number of reactor years. Let us assume that possible core melt-downs will occur randomly in time (i.e. according to a Poisson process), and also add some other homogeneity assumptions. It is then possible to deduce an expression for the upper confidence limit of the unknown "core melt-frequency" λ. Such a limit is larger than or equal to λ with a high fixed probability, the "confidence level" $(1-\alpha)$, often chosen as 95 or 99 per cent. If $N(t)$ is the number of the core melts up to t reactor years, the upper confidence level is

$$\lambda_1 = \frac{\chi_\alpha^2(2(N(t) + 1))}{2t}$$

where $\chi_\alpha^2(.)$ is found in standard statistical tables, see also Table 1.

TABLE 1

N(t)	$2\lambda_1 t$	
	$\alpha = 0.05$	$\alpha = 0.01$
0	5.99	9.21
1	0.49	13.28
2	12.59	16.81

We have now experienced about 600 reactor years without a core melt. That gives us an upper confidence bound of 3-5 per cent per reactor year.

Let us assume as an example that about 100 reactors are in use in the world during the next five years. If no core melt occurs during the coming five years the upper confidence limits based on statistical data will be of the same order as those of the Rasmussen report.

We can also illustrate the confidence limit as in Figure 2, where accident can represent a breakdown of any system, not only a core melt-down.

The example discussed here should not be taken too seriously when discussing the pros and cons of the Rasmussen report. We think however that it might be worthwhile to systematically compare statistical confidence limits with theoretical estimates. This should not only be done for the whole system but also for the various subsystems. In order to do this it will be necessary to collect data on incidents as is done by the Nuclear Plant Reliability Data System (NPRD) for U.S. reactors and by the Swedish Nuclear Power Inspectorate for Swedish Reactors (Andermo and Sundman, 1978). Much can also be learned from the safety analyses done for commercial airlines (Aeronautical Research Institute of Sweden, 1976-1977).

Concluding Remarks

Some of the difficulties encountered in trying to identify and estimate risks have been pointed out. The smoothing out of some

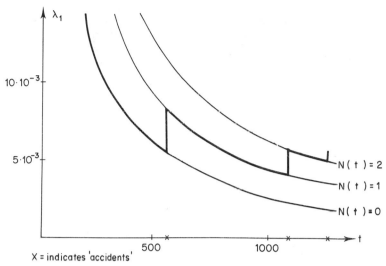

Fig. 2 Upper 95 per cent confidence bound λ, for the number of breakdowns per unit time as a function of t for different numbers of break-downs $N(t)$. The heavy line is for the case where the accidents occur at x-marks.

potential risks due to the mobility of people and/or actual policy diminishes the value of data registers. The large size and the rapid introduction of new technical systems give us fewer risk data. There are two ways to meet these difficulties. On a national basis we need to closely follow risk development even before catastrophes have occurred. And we also need international data on both incidents and accidents. This is not a plea for new organizations, but rather an expressed desire for good co-ordination in the collection and use of data.

Note Added in Proof

Let us consider break-downs at least as severe as that of Three-Mile-Island and let t stand for the number of reactor-years in the Western World. The x-mark about $t = 550$ indicates the TMI-accident and the upper confidence bound for such or worse accidents will be the thick line of figure 2 up to about $t = 600$. (The t-values are just crude estimates).

References

Aeronautical Research Institute of Sweden (1976-1977). "Hazard Briefs". (Brief description of aviation safety hazards from the HARP Incident Reporting System No. 1-9 FFA).
Andermo, E.G.L. and Sundman, B. (1978). "Report on Safety Related to Incidents and Reactor Trips". Swedish Nuclear

Power Inspectorate.

Rasmussen, N. (ed.) (1975). "The Reactor Safety Study". (Report WASH 1400). U.S. Nuclear Regulatory Commission.

Rowe, W.D. (1977). "An Anatomy of Risk". Wiley, New York.

KEEPING SCORE: AN ACTUARIAL APPROACH
TO ZERO-INFINITY DILEMMAS

Talbot Page

*Environmental Quality Laboratory,
California Institute of Technology,
Pasadena, California, U.S.A.*

A "zero-infinity dilemma", where there is presumed to be a very low ("nearly zero") probability of a very high ("nearly infinite") cost, is the least tractable yet most important type of risk problem associated with evaluating prospective energy alternatives. The term has often been used to describe the nature of the risk associated with nuclear power and many of the best known examples have been drawn from this field, but this type of risk problem is associated with the other major energy alternatives as well. The failure of the high-wall of a large hydro-electric dam to take just one example, poses another zero-infinity dilemma, with the risk of catastrophic destruction in the valley below it. Similarly, the risk of climatic change as a result of increased atmospheric carbon dioxide from fossil-fuel burning poses a latent zero-infinity problem.

The obvious difficulty in dealing with zero-infinity dilemmas is that there are little or no direct historical records to develop probability estimates based upon frequency of their occurrence. By way of illustration, in its summary, the Rasmussen report (1975) predicts that there is a one in five billion chance, per reactor year, of a catastrophic core melt-down. But the direct historical record of so many hundred reactor years with no melt-down is not greatly reassuring. This negative information, however welcome, is just as compatible with a one in thousand chance of catastrophic melt-down as with the one-in-five-billion Rasmussen estimate. The actuarial record, taken in simple terms of melt-downs per reactor year, is far too short to discriminate between the two. Nor do we want to wait for a long record of history of reactor years in order to develop an estimate of the probability of melt-down. On the contrary, we need the estimate of probability before a large scale programme is undertaken, in order to decide upon the safety features and the scale of the programme.

Decision Trees and Accountability

In spite of this somewhat pessimistic introduction, I believe
that an actuarial approach, broadly conceived, has something
constructive to say about the more accurate quantification of
risk even in the toughest case of the zero-infinity dilemma.
We may begin with a comment by William Stephenson (1976),
director of British Intelligence in World War II:

> "I can speak only of industrial 'accidents'. If you have
> access to insurance company files, you will see detailed
> studies of the weak point in any manufacturing process
> or mining procedure. Insurance companies stand to
> lose fortunes from an accident, and so they employ
> experts to figure out every possible way that things
> can go wrong. Their reports are guidebooks for saboteurs."

Two principal observations follow from this comment. First, the
comment suggests that decades ago insurance companies were using,
at least informally, an approach of error analysis which later
flowered in the enormous complexity of the Rasmussen report and
similar treatments of decision trees. From this observation it may
seem curious that as the procedures of risk assessment have become
greatly more sophisticated, the public's level of confidence in the risk
assessments, especially those related to nuclear power, appears to
have declined. This phenomenon has led some practitioners of risk
assessment to suggest that perhaps the public is irrational, not
understanding the techniques of risk assessment, in particular in
dealing with low probabilities. But, if this perceived phenomenon is a
real one, and public scepticism has actually increased, the second
observation suggests that this scepticism may in fact be rationally
based once the nature of the incentive structure is taken into
account.

In dealing with incentives relating to insurance, economists
have focused upon the behaviour of the insured party and the
problem of "moral hazard". As a second observation Stephenson
reminds us that we should also consider the incentive structure
upon the risk assessors themselves. In the normal practice of
insurance (which does not deal with societal zero-infinity dilemmas)
an insurance company can find out relatively quickly how accurately
it has assessed the probabilities of various accidents; moreover,
it stands to lose greatly if its assessments are inaccurate. The
situation is different when there are zero-infinity dilemmas.

Most zero-infinity dilemmas appear to share a number of
other characteristics besides the presumed low probability and
the potentially catastrophic outcome. Nine characteristics of
this type of risk have been discussed by Page (1978). In our discussion
here, the most important other characteristic is latency. For
example, in the greenhouse effect associated with fossil energy

sources, it takes many years for atmospheric carbon dioxide concentrations to build up and the projected climate changes to occur and decades to know whether or not there will be catastrophic effects following from our present decisions about fossil fuel. There are similarly long latency periods between the initiation of action and the manifestation of result from chemical carcinogens or radiation. Even such a dramatic event as a core melt-down is associated with a long latency in terms of the expected waiting time for its first occurrence. Latency means that there are few, if any, direct tests of the risk assessor's accuracy. Lack of a direct test means a lack of accountability, unless of course indirect tests can be devised.

Latency, by no means only associated with zero-infinity dilemmas, explains a common pattern of new industries. In the first stage, it is to the advantage of the industry's development to understate the risks and other costs associated with it. This incentive applies to all types of risks and costs but with special force to risks of the zero-infinity character because of the greater time-lag before actual knowledge of costs and risks is acquired. Certain alternatives, for example microwave transmission from satellite collection of solar energy, appear to be in this first stage. I am not suggesting here that the current assessments of risk and cost are in fact underassessments, but that at early stages of a new industry there are incentives, stemming from latencies of adverse effects, for such underassessment.

After a few decades, if several of the more optimistic predictions have proved false, a generalized scepticism is likely to set in, articulated in large part as observations about incentives upon the risk and cost assessors. It appears that the nuclear industry has moved into this second stage. One often hears such statements as "I have little faith in so-and-so's estimate, he consults for the industry", or "Of course he assesses the risks very low, he is a nuclear engineer and his career is at stake". While such *ad hominem* interpretations are no doubt galling to the risk assessors involved, protestations of objectivity have little impact at this stage. In the second stage the principal concern is establishing, or re-establishing, the credibility of the risk and cost-assessors. Again I wish to emphasize that there is no absolute connexion between an incentive and an outcome. But the public is right to pay attention to incentives.

Careful attention to institutional design and the role of incentives in it is beneficial for the following reasons: it increases the accountability of the risk assessors, and by developing more explicit direct and indirect tests of their predictions, is the most constructive way of increasing public confidence in the entire assessment process; at the same time, better assessments of risk may be obtained. Separating safety regulation from development in the U.S. Atomic Energy Commission (AEC), is an example of attention to incentives.

Keeping Score

In sporting events score is kept at two levels. At one level keeping score is a measure of the individual's or team's effort or success of effort, for example the number of goals scored in the first ten minutes of a basketball game. On the second level score is kept of the wins and losses of an individual in competition with others. Even where there is no formalized competition, score as a measure of effort is kept, for example in solo golf or in first ascents in mountain climbing, and this measure of effort is a requisite for scoring wins and losses in a formal, competitive setting. Still it is the scoring of the wins and losses that counts in most sports.

In the field of risk assessment, we are familiar with keeping score as a measure of the assessor's efforts, but less familiar with scoring the wins and losses of one assessor in direct competition with another. And indeed many may consider this an undesirable direction to go in, believing that the "game" of risk assessment is already too adversarial, especially in the area of nuclear energy. However, it is worth noting that the actuarial approach, in the setting of insurance industries, is directly competitive. One company competes against another, and the market keeps score of wins and losses.

In the area of energy, risk-assessment, especially for zero-infinity dilemmas, is mainly a government sponsored activity. But even if governments wished to establish competitive tests of performance, in some sense analogous to market tests, the task is made difficult by latency and the infrequency of observation of rare events. There are occasions, however, where latency is not a problem or where indirect tests could be constructed. It is interesting to speculate upon how a formalized competitive approach might go.

Consider the partly hypothetical example of the SNAP reactor, which was developed as a power plant for space vehicles. Before the satellite was launched the AEC predicted a very low probability of failure. An assessment of a 10^{-8} probability of failure appeared, although this was probably an informal and not official estimate. During the same period sceptics assessed the risk of failure to be much higher. While I doubt that a numerical estimate was produced that reflected the consensus view of the sceptics, or official estimate of a known organization of sceptics, such as the Natural Resources Defence Council, for the sake of illustration we may attribute an assessment of 10^{-4} probability of failure to NRDC.

We can imagine that a game is being played, with the immediate object of increasing one's credibility. We begin the game with an initial distribution of credibility among the players. On the basis of the relative sizes of the organizations and the numbers of experts in each, suppose we assign an initial credibility measure of .99 to the AEC and 0.01 to NRDC. The players gain or lose credibility depending on their risk assessments and the outcome of the experiment, in this case the launching of the satellite. We define two states:

Θ_{AEC} The AEC is accurate in its risk assessment

Θ_{NRDC} NRDC is accurate in its risk assessment;

the two risk assessments for the SNAP reactor failure (F):

$$P(F \mid \Theta_{AEC}) = 10^{-8}$$
$$P(F \mid \Theta_{NRDC}) = 10^{-4}$$

When there are many experiments, or rounds, we can write

$p^t(F \mid \Theta_{AEC})$ AEC assessment of probability of failure in round t;

$p^t(F \mid \Theta_{NRDC})$ NRDC assessment of probability of failure in round t;

$p^t(S \mid \Theta_{AEC})$ AEC assessment of probability of success in round t;

$p^t(S \mid \Theta_{NRDC})$ NRDC assessment of probability of success in round t.

We start with the two initial measures of credibility:

1)
$$C(0,AEC) = 0.99;$$
$$C(0,NRDC) = 0.01;$$

these initial values correspond to poll positions in a horse race.

We can define the credibility measures in round t by a formula corresponding to Bayes' Theorem:

If round t turns out to be a failure

2) $$C(t,AEC) = \frac{p^t(F \mid \Theta_{AEC})C(t-1,AEC)}{p^t(F \mid \Theta_{AEC})C(t-1,AEC) + p^t(F \mid \Theta_{NRDC})C(t-1,NRDC)}$$

And if round t turns out to be a success

$$C(t,AEC) = \frac{p^t(S \mid \Theta_{AEC})C(t-1,AEC)}{p^t(S \mid \Theta_{AEC})C(t-1,AEC) + p^t(S \mid \Theta_{NRDC})C(t-1,NRDC)}$$

and

$$C(t, NRDC) = 1 - C(t,AEC)$$

In the Second Round we would start with credibility measures

(C(1,AEC),C(1,NRDC)), and the contested risk assessments for some other event, which could be a second try for the SNAP reactor or a quite different event whose rarity is at issue between the AEC and NRDC, but which can be observed and scored.   Then we would compute the Second Round credibilities from (1) with t=2. The initial credibility measure (C(0,AEC), C(0,NRDC) is the same as the prior distribution in decision analysis; and the First Round credibility measure (C(1,AEC),C(1,NRDC)) is the same as the posterior distribution. It would not be worth recasting the notation from that in standard decision theory, except for the slight extension from the usual interpretation – it is assumed that θ_{AEC} refers to an entire technique and capacity for risk analysis, and thus the credibility measure describing the AEC assessment of a SNAP failure also describes AEC assessments of other risks (see also Howard *et al.*, 1972).[1]

In the case of the SNAP reactor, the experiment was undertaken and unfortunately the reactor failed upon its first try.   Direct computation from Bayes' Theorem shows that the credibility measure for the AEC falls from 0.99 to 0.01 and the measure for NRDC rises correspondingly. Such an enormous fall in credibility is startling and it raises a number of interesting issues. First of all, it illustrates that the initial seeding of players, which is done on the basis of outside judgement, can be quickly overcome by an actual "track-record." On the contrary we might ask whether a credibility measure should be defined to be capable of such volatility? This is really a question of the applicability of standard decision theory to low probability events, as it has nothing to do with our assumption of extension, which says that the credibility measure applies to the entire risk assessment capacity, not just one prediction, and allows us to go from one round to the next. This question has to do with the relative weights to be given to the prior and the new evidence, which is a basic question in decision analysis.

Without deeper discussion of this last question, we may note that the volatility is asymmetric. If the SNAP reactor had succeeded NRDC'S credibility measure would have fallen, but its percentage change would have been much smaller.   At least it can be said that the asymmetry is in the right direction. When both parties agree that a failure is rare, there is more information in a failure than in a success (as to the accuracy of the risk assessments).

---

[1] In this discussion on the decision to seed hurricanes, the prior and posterior distributions can be interpreted as original and revised credibility measures on the hypotheses put forward by differing scientists. In Howard *et al.*, there is no assumption of an extended and roughly constant capacity of risk assessment, and there is just one round.

If we look at this credibility "game" as a sequential process lasting hundreds or thousands of rounds, where not all the probability assessments have to be on rare events, we may ask questions of convergence. Suppose that the AEC were indeed more accurate in its risk assessment than NRDC. How long would it take to show up in its credibility measure? In a long sequence of rounds, how much does the initial posting or seeding of players matter? Given that the object of the game is to maximize one's own credibility over a sequence of rounds, is a player better off publishing his own "sincere" estimates of risk, or other numbers depending on what he thinks the other player is going to do? In other words, are there dominant strategies and if so, do they produce unbiased estimates of risk?

If credibility games could be developed with sincere dominant strategies, they could provide powerful incentives toward more accurate risk assessment. The incentive would apply to both players, so that one would expect a convergence of risk assesssments, a welcome phenomenon. An important purpose of establishing a formal game of the sort sketched above is to provide for near term accountability for both players, for assessments and statements about their assessments. At the present time it is unclear to the outsider how seriously to take the failure of the SNAP reactor, for example. It is possible that the AEC has made hundreds of thousands of predictions and NRDC has seized upon the unlucky one. On the other hand it is possible that the AEC got away with overly optimistic predictions by not making them formal and then keeping quiet about the failures. Without keeping score on the wins and losses there is no way to tell.

Even though at present there is no accountability of the risk assessors along the lines of the formal game outlined above, to some extent this game informally exists already (and in a marred form). Agencies lose credibility when they assign very low probabilities to events which then happen, especially when critics have assigned larger probabilities. Agencies gain credibility in the reverse situation. Thus it would be useful to analyse the winning strategies in such games to see what incentives are on the players, even in the informal situation.

As mentioned earlier, even without establishing a formal competition and scoring the wins and losses of the players in comparison with each other, it is possible to keep score on the individual level, of the solo player. When we are dealing with final events, such as the success or failure of a reactor, with very low assessed probabilities of failure, just one or two failures casts doubt upon the assessment process, unless there are hundreds of millions of low probability assessments being made.

A comparison can be made with the problem of assessing the probability of a perfect bridge hand (where one player gets all clubs, another all hearts, etc.). Textbook exercises, based upon combinatorial principles, yield an extremely low estimate of probability. The estimate is so low that a single occurrence of a perfect hand casts doubt on the probability estimate. And as perfect hands are reported every few years, the assessment technique itself has been reconsidered. Upon review it appears that some perfect hands are the result of gags (which would correspond to Stephenson's sabotage). Moreover, it appears that when sabotage is prevented, as in tournaments, the probability of a perfect hand is still too high to be explained by a combinatorial calculation. It appears much more likely that perfect hands result from a "common mode failure" – that is, the probabilities of one intermediate event are conditional upon another intermediate event. New decks come in perfect hands and shuffling is not entirely random, so that dealing is not made up of independent events. Thus even the single occurrence of a presumed very rare event can lead to the reassessment of the risk assessment technique and the adoption of more realistic techniques.

The evaluation of risk assessment is made more difficult by the fact that many of the gambles being assessed are not repeatable experiments. They are assessments of gambles about the state of nature, which may be one way or another. Each gamble is assessed at its own probability, typically a very low one for the kinds of problems we are considering, and then scored 0 or 1, depending on whether the benign or adverse hypothesis about the state of nature turns out to be true. It is not possible to tell whether or not a single assessment of the likelihood of a particular hypothesis is realistic or not. However, with the assumption of extension (that an agency's assessment technique is consistently accurate or inaccurate) it is still possible to score an overall effort.

Suppose that an agency were totally accurate in its risk-assesments and its estimate of the risk of failure (or adverse hypothesis) of the ith gamble is p_i. Then the sum of N such Bernoulli random variables has expectation

$$\sum_{i=1}^{N} p_i$$

and variance

$$\sum_i p_i(1-p_i) .$$

This binomial-like random variable (the p_i are different) allows us to score the aggregate effort of the risk assessor and ask how likely it is that the assessor has been accurate.

Because of latency, there is limited opportunity to develop this type of scorecard for a solo player, if we insist upon evaluating probability estimates of final events, such as core melt-downs.

However, this technique applies to the scoring of intermediate events and partial chains in decision trees. In fact it may apply more suitably to the evaluation of probability assessments of intermediate events than final events. Not only are there more intermediate events to evaluate, often with less of a latency problem, but also evaluation of intermediate events lends itself to testing assumptions of independence. Common mode failures take place in the intermediate chains of events more directly than across final events. As the binomial-like distribution assumes independence of the underlying Bernoulli random variables, this assumption is tested simultaneously with the assumption of accuracy of the individual assessment of the p_i.[2]

The binomial-like distribution provides a way of testing the aggregative performance of the risk assessor. Depending on the types of assessments being evaluated it can also test the assessor's allowance for, or treatment of, the possibilities of human error (e.g. the initiation of the Brown's Ferry accident) or his own tunnel vision (leaving out paths or focusing on the wrong ones).

Conclusion

Interpreted narrowly as the method of calculating probabilities of final events by their historical frequency, the actuarial approach has limited application for risk analysis in the energy area, especially where there are low probabilities of catastrophes, and often latencies. Interpreted broadly as the approach to risk estimation

---

[2] The assumption of extension — that the accuracy ability of an individual risk assessor is roughly constant across experiments — appears to be the underlying principle in a series of insightful experiments on the evaluation of risk assessors by Alpert and Raiffa and by Tversky and Kahneman. These investigators used a list of "almanac" questions such as: "How many foreign cars were imported into the United States in 1968? For each question the test subjects were asked to make an estimate so high that they would believe there was only a one per cent probability that the true answer would exceed their estimate, and similarly for a low estimate. Thus the subjects constructed rare events — being outside their confidence intervals. Although there is no way of telling if a single estimate is accurate, under the assumption that the assessors were consistently accurate one would expect the factual answers to be outside the constructed intervals about two per cent of the time, over a number of trials. The experiments suggested that the subjects were not consistently accurate assessors, because the factual answers were outside the constructed intervals 40 to 50 per cent of the time. These experiments are described in Slovic et al., (1974).

found in insurance markets, the actuarial approach has much to offer. The actuarial approach, in the context of market competition, uses the historical record to reward the companies with accurate risk assessors and starve the others.   The actuarial approach directs us to increase the accountability of the risk assessors in government sponsored assessments and keep score of the success of effort of the individual risk assessors.   The approach suggests an analysis of the incentive structure on the assessors themselves, and of the strategies of the (informal) games the assessors are placed into.   The goals of the actuarial approach, broadly conceived are:

1) to achieve better point estimates of probabilities;
2) to achieve a better understanding of the credibility or subjective confidence interval to be associated with the point estimates;
3) to understand better the incentives for under- or overassessment;
4) to increase the level of confidence of the public in the entire assessment process, through explicit evaluation of performance.

References

Howard, R.A., Matheson, J.E. and North, D.W. (1972).   The decision to seed Hurricanes. *Science* **176,** 1191-1202.

Page, Talbot (1978).   A generic view of toxic chemicals and similar risks. *Ecology Law Quarterly,* symposium issue.

Rasmussen, N. (ed.). (1975).   Reactor Safety Study: an Assessment of Accident Risks in US Commercial Nuclear Power Plants. US Nuclear Regulatory Commission.

Slovic, P., Kunreuther, H. and White, G.F. (1974). Decision processes, rationality, and adjustment to natural hazards. *In* "Natural Hazards: Local, National, and Global" (G. White, ed.).

Stevenson, W. (1976).   "A Man Called Intrepid."   Ballantine Books, New York.

INSURANCE AND LOW-PROBABILITY RISKS

H. Bohman

*Skandia Insurance Company,
Stockholm, Sweden.*

In the previous discussion, "Keeping Score," Dr. Page presents an example where two assessors have different initial credibilities and give different estimates of the probability of a certain event occurring (the failure of the power plant of a space vehicle). If we apply the Bayesian method we get a drastic change in credibility should the failure actually take place. It seems to me that this example is an excellent warning that probability theory as applied to small probabilities might produce very silly results from a practical point of view. Many references are made by Page to actuarial methods and the actuarial approach. I will therefore describe briefly how the insurance industry looks upon the problem of small probabilities.

The insurance business is an application of statistics. You have to rely on statistical evidence to produce equitable premiums for different lines of business. It must, however, be remembered that the basic idea behind the insurance business is to equalize the cost of insurance between similar risks. In order to operate an insurance business in a safe and reliable way, the whole insurance portfolio must be composed of homogeneous sub-groups of insurances which are large enough to permit a reliable estimation of the costs of risks and how these costs can be distributed among the policyholders in an equitable way.

In the insurance business, there is a need for a precise definition of risk and an accurate way to measure risks. For this purpose we use the frequency per year and per unit of observation, of the event which creates the risk in question. In life insurance the risk is that the insured client dies. In fire insurance the risk is that the house catches fire. In motor third-party liabilities the risk is some harm caused to somebody by a person driving a car.

This way of measuring risks is very practical and useful. I want especially to emphasize that this standard method makes

possible relevant comparisons between different risks. A male
Swede at age 60 has a rate of mortality equal to 0.01. It happens
that the rate of mortality in scheduled air traffic is approximately
the same. Approximately 1 per cent of such aircraft crash each
year. As a consequence, you can state that as long as you travel
in a schedule aircraft you are exposed to the same risk of dying
as a 60 year old male Swede is exposed to constantly. This way
of describing the risk of flying is much more illustrative than
the traditional comparison of air-travel with steamers and railways.

The frequencies as described above are very important for
the insurance industry because the calculation of premiums is
based on these frequencies. Generally speaking, it can be said
that the insurance industry trades in risks where the frequency
is at least equal to 0.001. If the frequencies are smaller, it will
always be difficult to arrive at reliable statistical estimates
of the frequencies in question. These difficulties are reflected
in constant disagreements between the insurance companies
and their clients. The clients think that the insurance companies
use an estimate of the frequency which is too high, and the insurance
companies, lacking reliable statistical data, have to apply estimates
which they believe to be on the safe side.

If the frequencies are larger, the statistical estimates become
safer and there is in practice no problem in convincing clients
that their premiums are reasonable.

If the frequencies are very small the premiums will also be
very small compared with the risk the insurance company has
to face, namely, to pay a large insurance amount. From experience,
the insurance industry knows that there must be a reasonable
relationship between the insurance amount and the premium
paid.

Summarizing the experience of the insurance industry, it
can be said that insurance risks with very small frequencies create
such a problem to the industry that it rather tries to avoid insuring
against such risks. There must be statistical evidence to prove
that the premiums are equitable, and such evidence is very difficult
to get in the case of really small frequencies.

Does this mean that the insurance industry cannot or will
not cover risks where the frequencies are very small? The answer
is no. Exceptionally the companies accept such risks, but it is
then meaningless to ask for an equitable premium or demand
that the insurance company prove that the premium is equitable.
This is simply not possible. The premiums will be a qualified
guess and it is up to the client to accept or reject the offer for
insurance protection. If he is really anxious, he is likely to accept
the insurance policy even if he does not consider the premium
equitable. When he does so, however, he is acting not upon mathematical
and statistical evidence but rather on his feelings, his fear and
his intuition: and the same goes for the insurance company when
it sets the premium.

PART 3
OBJECTIVE AND PERCEIVED RISK

INTRODUCTION

The opening paper by Rowe in Part 1 outlined the problem of the divergence between 'perceived' and 'objective' risk and this was further developed (in Part 1) by Kasper. In the ensuing Part, these two basic approaches to risk estimation and evaluation are elaborated further still. The first two papers, on the management of risk from ionizing radiations and genotoxic chemicals respectively show how far it is possible to go at present by using purely objective data obtained by applying classical scientific principles and measurement techniques. By contrast, the third and final paper in this Part sets out to analyse logically and systematically exactly what subjective factors, value-scales and anxieties people respond to when making decisions within risk-situation. Such subjective perception cannot ethically be ignored by any humane society. Nor can the pragmatic politician fail to recognize its power, especially in those democracies where such issues enter into election campaigns.

Efforts to estimate and control ionizing radiation have been evolving over a fifty-year period and serve as a model for other genotoxic agents such as radio-mimetic chemicals. The first paper by Bo Lindell, who is also the Chairman of the International Commission on Radiological Protection (ICRP), not only traces the history of radiation-protection practice, but describes in detail the present approach. The need for different treatments for acute (non-stochastic) and chronic (stochastic) exposures was recognized early and the difficulties of establishing standards in the absence of thresholds for stochastic exposure have been well understood. Standards for radiation protection – the means by which levels of acceptable risk are established in this area, have evolved from levels set to protect against acute exposure, i.e. "tolerance limits": to a qualitative concept of "as low as practicable", to the present process of assigning a monetary value to radiation detriment in order to establish quantitative values. This value, which is established under the direction of the national authorities responsible for radiation protection, is concerned with the collective risk to exposed populations.

The process has three distinct steps: justification; optimization; and the establishment of individual dose limits and allocation of individual doses among identified sources. Justification involves a balancing of the benefits and costs of a new activity, including radiation detriment and protection costs. It is primarily a balance made at the policy level, and has the problems associated with cost-benefit analysis inherent in its use. However, the separation of radiation detriment and protection costs from the remaining factors makes these visible. Optimization is directed at determining how much should be spent to reduce the collective (population) dose by assigning a monetary value to radiation detriment as a limit. These two steps look at the total dose commitment over various time scales. In this way they examine the problem of genetic and somatic risks to future generations explicitly. Dose limits are used to assure that inequitable risks to individuals are kept low. The system is complex, detailed and explicit, but still is based upon scientific value-judgements about the uncertainties in risk estimation and social value judgements in risk evaluation. However, the value judgements are visibly identified, and the social judgements left to national policy makers.

The second paper by Lars Ehrenberg and Göran Löfroth examines the problems of objectively estimating the risks to health from physical and chemical genotoxic agents. They point out that the difficulties lie in identifying and quantifying the different specific effects from known and unknown genotoxic agents. As with ionizing radiations, linear extrapolation to low doses gives, at the present state of our knowledge, the most probable shape of the dose-response relationship for chemicals promoting mutations and cancers. Among the chemicals of greatest interest here are the polycyclic aromatic hydrocarbons, emitted during the combustion of fossil-fuels. Genetic risk from such compounds is related back to the better understood genetic risk from ionizing radiations via the radiation-risk-equivalent ("rad-equivalent" or "Sievert-equivalent").

The final paper addresses "perception" of risk directly. Paul Slovic, Sarah Lichtenstein and Baruch Fischhoff accomplish this by asking different groups about risks, particularly risks from nuclear energy. This interrogative approach to studying perceptions of risk is relatively new and has also been developed over the past few years by Otway and his co-workers at IIASA in Vienna and by Sjoberg in Sweden. But the technique is still in its infancy. Many of the efforts have examined issues broader than nuclear energy, but the apparent divergence of perceived risks from objective estimates in the nuclear area provides a unique array of conditions for understanding why this divergence exists.

The authors conclude that the basis for the disparity between perceived and estimated risks is a fundamental cognitive mechanism which leads people to use imaginability and memorability as the basis for estimating the probability, frequency and magnitude of events. Nuclear hazards are generally characterized by the "zero-infinity dilemma" and latency (as described by Page in Part 2), and are not amenable to empirical testing and verification.

The nature and coverage by the news media of nuclear hazards make them particularly memorable and imaginable, leading to wide disparity between "experts" and significant portions of the public, without the opportunity for testing.

A most interesting conclusion is that Page, by studying actuarial risks, and Slovic *et al.*, from direct analysis of risk perception, arrive at the same point. They both recommend the development of mechanisms and institutions to provide credible estimation, testing and accountability for such risks.

DEVELOPMENT OF STANDARDS RELATED TO RECEIVED-DOSE: THE RADIONUCLIDE CASE

Bo Lindell

Swedish National Institute of Radiation Protection, Stockholm, Sweden.

In the radiation field, the International Commission on Radiological Protection (ICRP), operating since 1928, issues basic recommendations on protection. These recommendations form the basis for regulations on radiation dose-limitation in most countries and are recognized by international organizations such as IAEA,[1] ILO,[2] OECD/NEA,[3] UNEP[4] and WHO.[5]

Up to the 1950s, it was generally believed that radiation harm could not be caused unless the radiation dose exceeded a certain *threshold* value. This was, in fact, true for the *non-stochastic* effects (tissue destruction, e.g. destruction of the bloodforming organs, due to cell death), which were the alarming consequences of unsafe radiation work in the early part of the century.

The radiation dose, or strictly the *absorbed dose (D)* of radiation is the energy eventually absorbed per unit mass of the irradiated body. The present SI unit is joule/kg, for which the *gray* (Gy) is accepted as a special name. The old unit was the *rad* (1 rad = 0.01 joule/kg).

In the first half of this century, the aim of radiation protection was to ensure that no individual received a radiation dose which exceeded the threshold value for harmful effects. For this purpose an individual dose limit was early introduced. This was often referred to as the "tolerance dose", because it was set with such a wide safety margin below the dose threshold, that the body could

---

[1] International Atomic Energy Agency;
[2] International Labour Organization;
[3] Organization for Economic Co-operation and Development/Nuclear Energy Agency;
[4] United Nations Environment Programme;
[5] World Health Organization

easily tolerate the irradiation up to that limit.

In the ICRP terminology this dose limit became known as the "maximum permissible dose" (MPD), the recommendation being that no national authority should issue regulations that would permit higher doses. The ICRP MPD-value for occupational exposure to X-rays and to gamma radiation from radium corresponded to 300 mrad per week until the mid-50s, when the limit was reduced to 100 mrad per week for the most sensitive organs (the gonads and the blood-forming organs).

Because of the belief that the tolerance dose was a safe dose, it was recommended that a weekly dose of that magnitude could be considered "for purposes of planning and design" and that exposures up to the dose limit were quite acceptable. Since such high exposures would only occur for radiation workers, no dose limits were originally given for members of the public. The special case of radiation exposures of patients in radiation therapy and in roentgenological examinations was considered to fall within the responsibility of the medical practitioner who would limit the doses in the individual cases as he felt was in the best interest of the patient. In radiotherapy, for example, it was obviously impossible to destroy a tumour by radiation without simultaneously causing some harm to surrounding normal tissues.

In the mid-50s, the situation changed. The radiation protection efforts had in most developed countries resulted in a situation where doses in excess of the dose limits were rare. Non-stochastic effects would then only occur in cases of accidental high exposures or as a result of gross negligence in following regulations and instructions. The interest then turned to *stochastic* effects such as hereditary harm and cancer, including leukaemia.

While the non-stochastic effects would only occur above a dose threshold and then increase in severity with increasing dose, the severity of stochastic effects would be essentially independent of dose but the effects might occur with a probability which would increase with the dose without any obvious threshold. The true form of the dose-response relationship for the stochastic effects, however, was not known and most biologists doubted that very small doses would cause any such effects.

In spite of these doubts, many radiation protection recommendations were formulated on the basis of a possible direct proportionality between dose and response, for the sake of cautiousness. With this cautious approach, it was not possible to guarantee that doses below the dose limits were entirely safe. The ICRP (1955) warned that

> "Whilst the values proposed for maximum permissible doses are such as to involve a risk which is small compared to the other hazards of life, nevertheless, in view of the incomplete evidence on which the values are based, coupled with the knowledge that certain radiation effects are

irreversible and cumulative, it is strongly recommended that every effort be made to reduce exposure to all types of ionizing radiation to the lowest possible level."

In 1958, this wording was changed to a recommendation that all doses be kept *"as low as practicable"* (ICRP, 1958), and in 1965 the phrase became that all doses be kept *"as low as is readily achievable, economic and social considerations being taken into account"* (ICRP, 1966). In 1973, it was suggested that the word "readily" should be replaced by "reasonably" (ICRP, 1973).

In the most recent basic ICRP recommendations (ICRP, 1977a), this recommendation has been given even more emphasis, backed by biological and epidemiological evidence provided by the United Nations Scientific Committee on the Effects of Atomic Radiation (UNSCEAR), which published its most recent report to the UN General Assembly in 1977.

The gradual shift in emphasis from the 1950s when the non-threshold dose-response relationship was first considered but not much trusted, until today's situation when it is still considered but believed to be quite likely, has caused a corresponding shift in radiation protection policies (Lindell, 1978). With no dose considered a safe dose, radiation protection norms have developed from simple rules for avoiding high occupational exposures to rather complicated methods for eliminating even minute exposures of the public, when this is believed to be reasonably achievable.

The Dose-Response Relationship

Even though it was first believed that the assumption of direct proportionality between dose and response grossly overestimated the true risk at low doses, it was early realized that evidence of the true dose-response relationship at low doses could hardly be expected from epidemiological observations on man, considering the small magnitude of the risk.

Today, it is generally agreed that the dose-response relationship is very likely to be linear over the limited region of small dose increments above the background exposure that is of interest in radiation protection. The slope of the dose-response curve at low doses, however, is not known. It is often taken to be equal to the slope of the straight line extrapolating the observed response at high doses towards the origin. It is still debated whether this provides an over-estimate or an underestimate of the true risk, but it is generally agreed that the error cannot be a large factor (Swedish Energy Commission, 1978).

It should be noted that the postulation of a linear, non-threshold dose-response relationship does not necessarily depend upon the reliability of the biological assumption. In many practical cases, radiation protection deals with the limitation of doses which are accumulated from a large number of sources, each of which contri-

butes but a little.   In the practical administration of radiation
protection it would be very difficult to distribute the responsibility
for the final result unequally, as would in fact be the case if, with a
non-linear dose-response relationship, the last exposures carried a
higher risk per unit dose than the previous ones.   In controlling the
situation, including the future development, the authorities would
have to work on averages and postulate an equally shared
responsibility for the final result, which amounts to the same as
postulating a linear dose-response relationship.

The important practical consequence of the assumption of a linear
dose-response relationship without a threshold is that it makes it
possible to calculate consequences on the basis of average doses,
irrespective of the actual dose distribution (provided that the threshold for non-stochastic effects has not been exceeded in any
individual case).   For example, the mathematical expectation of the
number of cases of stochastic effects in an exposed population would
be proportional to the product of the number of individuals and their
average dose (as will be seen in the following, this product is called
the *collective dose*).

The Effective Dose Equivalent

One and the same absorbed dose (*D*) from different types of radiation
may have different biological effects (different degrees of severity of
non-stochastic effects, different probability of stochastic effects),
because of different relative biological effectiveness (RBE) at different radiation qualities.

In Reports 19 and 25 (1971, 1976) of the International Commission
on Radiation Units and Measurements (ICRU), a special quantity, the
dose equivalent (*H*) is defined by the equation

$$H = Q \, N \, D$$

where *D* is the absorbed dose, *Q* is the *quality factor* given by ICRP
as a function of the collision stopping power of water, and *N* is the
product of any further factors which might subsequently be recommended by ICRP for this purpose (so far *N*=1).

Since *Q* and *N* are dimensionless, the dose equivalent has the same
dimension as the absorbed dose and may therefore be expressed in the
same units.   In order to avoid confusion, however, a special name, the
sievert (Sv), has been introduced by ICRP and ICRU for this unit (1
Sv= 1 Gy = 1 J/kg).   The sievert replaces the former unit, the *rem*,
which was related to the rad by the above equation (1 Sv = 100 rem).

By expressing dose limits as limits of the dose equivalent rather
than of the absorbed dose, the numerical limits can be given more
simply, and independent of the type of radiation.

In previous recommendations, ICRP gave dose equivalent limits
for a number of organs and tissues.   This former system of dose
limits, however, was criticized as being inconsistent.   For example,

the dose equivalent limit for the gonads was 5 rem in a year, irrespective of whether other tissues were irradiated or not, the dose equivalent limit for the whole body being also 5 rem in a year.

In its new Publication 26 (1977a), ICRP introduces the recommendation that the sum of the weighted individual organ dose equivalents (H_T) should be subject to the dose equivalent limit (H_L) that would apply in the case of uniform whole body exposure:

$$\sum_T w_T H_T < H_L$$

The weighting factors w_T recommended by ICRP for the various organs and tissues (T) have been chosen to be proportional to the assumed risk (r_T) per unit dose equivalent of the occurrence of a serious stochastic effect (death from cancer, serious hereditary defects) after the exposure, the genetic effect being assessed for the first two generations. The average dose equivalent to produce one case of such serious stochastic effect is w_T/r_T. Irrespective of the dose distribution within the body, one and the same value of the sum of the weighted organ dose equivalents is always related to the same risk of serious stochastic effects, since $w_T/r_T = 60$ Sv in all cases. It follows that 60 Sv is the dose equivalent that, in the case of uniform whole body exposure, would on the average be needed to cause one case of serious stochastic effect

At a meeting in Stockholm in May, 1978, ICRP (1978) decided to name the quantity which is defined by the above sum the *effective dose equivalent*, as first suggested by Jacobi (1974). This quantity (H_E) is therefore defined as

$$H_E = \sum_T w_T H_T$$

Irrespective of the dose distribution in the body, the effective dose equivalent can be taken as the representative dose equivalent, to which dose limits and risk factors may be applied. It must be recognized, however, that the total risk factor of $1.65.10^{-2} Sv^{-1}$ (for details see ICRP, 1977b) is only applicable to an *average* individual and that higher or lower risk factors would in fact apply to other individuals, depending upon age and sex. The variations are, however, not large enough to warrant any higher precision in the weighting factors. In ICRP Publication 27 (1977b), it has been shown that in most practical situations the use of the effective dose equivalent and a risk factor of $1.65 \times 10^{-2} Sv^{-1}$ would overestimate the risk if the age- and sex-related dose-response estimates of ICRP (1977a) and UNSCEAR (1977) are valid. The risk would, however, be underestimated for young women and the risk factor does not include hereditary harm expressed after the first two generations. This late genetic harm has been estimated to be of about the same magnitude as that expressed in the first two generations (ICRP, 1977a).

The Basic Principles of Radiation Protection

ICRP (1977a) refers to the following three basic principles of radiation protection:

1) The practice causing radiation exposure should be justified on an overall cost-benefit basis;
2) Radiation protection should be optimized so that all doses are kept as low as is reasonably achievable; and
3) All individual doses should be kept low enough to guarantee appropriate individual protection.

Justification of the Practice

In limiting releases of radioactive substances into the environment, a political decision as regards the fulfilment of the first requirement will be needed before authorized release limits can be given by the authorities. This decision is taken at the appropriate political level, for example by the government in cases of great consequence (e.g. the acceptance of nuclear power) but perhaps by the radiation protection authority in minor cases, after consultation with those who can assess the benefit of the practice.

It must be remembered that the justification assessment is not a scientific but a political decision. A number of intangibles have to be considered and it is not realistic to believe that an answer can be derived merely from numerical calculations. However, the total radiation *detriment* is one factor for consideration.

The radiation detriment (G) is a quantity introduced by ICRP (1977a). It is the mathematical expectation of future harm (some of which may occur first after long latency periods) incurred by a radiation exposure, taking into account not only the probability of each type of deleterious effect but also the severity of the effect. In the simplified case when N individuals all receive the same effective dose equivalent H_E, the detriment will be

$$G \sim g \, N \, H_E = g \, S_E$$

where g is the total individual detriment per unit dose and $S_E = N \, H_E$ is the *collective effective dose equivalent*. This is not an exact expression of the total detriment, since the definition of H_E does not include weighting of the gonad dose for harm expressed after the first two generations, but this is a small deviation, considering other large uncertainties in the assessment.

In justification assessments it is often helpful to express the detriment of a given practice per unit of the practice (e.g. per year of operation or per terawatt hour (TWH) of electrical energy produced).   It is then a first step to assess the total collective effective dose equivalent per unit of practice.   Some of the collective dose may be caused by exposures long after the practice, due to long-lasting radioactive contamination of the environment. It is then the collective dose *commitment* (Lindell, 1979) that is the quantity of interest. The collective effective dose equivalent commitment (S_E^C) is defined as the infinite time integral of the collective effective dose equivalent rate:

$$S_E^C = \int_0^\infty \dot{S}_E \, dt$$

In the case of environmental pollution with very long-lived radionuclides (e.g. carbon-14, iodine-129, uranium-238 and plutonium-239), the collective effective dose equivalent commitment and hence the detriment $(G = gS_E^C)$ will relate to very long time periods $(10^3-10^{10}$ years). It is evident that the assessment of the collective dose rate in such a distant future is extremely uncertain, but on the basis of maximizing assumptions (e.g. that the substance will become unavailable in the future only because of radioactive decay) upper values may be assessed.   In addition to the uncertainty of the radioecological assumptions, which may be chosen conservatively, the assessment requires the assumption of the existence of a human population of a postulated size over such long periods of time. It is often assumed, for the purpose of such assessments and in spite of the probable lack of resources to maintain such a large population, that the world population will be of constant size equal to 10^{10} persons over all future time (Beninson, 1978a).

Because of these uncertainties, the political decision makers should not be presented with only a number giving the total detriment or the total collective dose commitment, but also information on how this number is expected to be accumulated over time.

The expected accumulated harm $G(t)$, i.e. the part of the detriment that relates to the period up to a certain time t, may perhaps be assessed with some confidence for periods over hundreds or thousands of years. For longer periods, however, the uncertainties will blur the picture, as indicated in Figure 1 which illustrates the harm accumulation on the basis of the time scale for iodine-129.

It has been suggested[6] that future harm might be given less weight than present harm and that a discounting factor, e.g. e^{-kt} with a k-value of the order of 10^{-2} year $^{-1}$, should be introduced in the calculation of $G(t)$ or $S^C(t)$. This, again, would be a political

---

[6]The proposal of discounting future doses was early expressed by H.L. Gjørup at the Risø Research Centre in Denmark (personal communication).

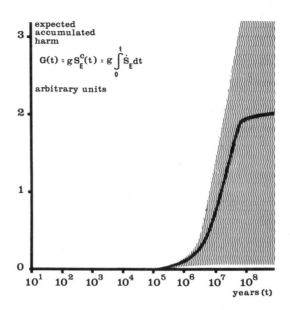

Fig. 1 Expected accumulated harm $G\ (t)$, with time-scale based on half-life of iodine-129. Shaded area represents uncertainty range.

rather than a scientific conclusion. It is difficult to see the ethical justification of discounting future harm. However, it might be justified to discount the weight of future doses in the justification assessment, not because they are future but because of the doubt that there will be a reality behind the estimates.

Dose Limitation

If a practice has been found justified by political decision, it remains for the competent national authority to give norms and regulations on radiation dose limitation, usually on the basis of the ICRP recommendations.

Because of the requirement that all individual doses should be kept as far below the ICRP dose limits as it is reasonably achievable, it is not possible to accept a situation only on the basis that monitoring data show that all individual doses fall below the dose limits. Nor is remedial action such as restriction of land uses, prohibition of the sale of contaminated food, etc. necessarily justified because the dose limits might be exceeded. The distinction between *action levels* (derived from risk-benefit considerations of the possible remedial actions) and operative, *authorized limits* for normal operations is important.

The old policy for authorized limits, now abandoned in most countries, was to use the recommended ICRP dose limits for purposes of planning and design, although usually with some margins of safety due to cautious assumptions on the relationship between practice rate and resulting radiation doses. The rate of release of radioactive substances that was expected to cause, under steady state conditions, radiation doses at the ICRP dose limits, was called "the capacity of the environment" to receive radioactive material from the particular source. Authorized release limits were usually set as a fraction of this release rate, but the situation was often considered acceptable as long as the "capacity of the environment" was not exceeded.

Today, authorized release limits are set on the basis of the most restrictive of the following considerations:

a) Optimization of protection;
b) Individual dose limitation for critical groups; and
c) Control of future *per caput* doses in large populations.

Optimization of Protection

Optimization of protection is based on the second of the three basic protection requirements mentioned earlier often called the "ALARA" principle, namely that all doses should be kept "as low as reasonably achievable, economic and social consequences taken into account" (ICRP, 1973; IAEA, in press).

Optimization of protection is achieved when the cost of additional protection would be greater than the resulting reduction in radiation detriment, while at the same time the possible reduction in cost of protection by reducing the protection efforts is less than the resulting increase in radiation detriment. At that point:

$$\Delta X = - \Delta G$$

if X is the cost of protection and G is the detriment resulting at that cost level G would have to be expressed in monetary terms in order to make the comparison possible. However, since the detriment is assumed to be proportional to the collective effective dose equivalent commitment S_E^C, the condition for optimized protection will be

$$\Delta X / \Delta S_E^C = - \alpha$$

where α is the detriment (in monetary terms) per unit collective effective dose equivalent commitment. Values for α of the order of 10^4-10^5 U.S. \$ per sievert have been suggested (ICRP, 1973; U.S. Nuclear Regulatory Commission, 1976).

Often there is no continuous correlation between the cost of protection (X) and the corresponding collective dose commitment S_E^C. In many practical cases, a number of different alternatives (e.g. different type of retention system for release reduction) may need to be considered for the selection of the one which gives the optimum result (Beninson, 1978b). A system B is then preferable to a system A if:

$$\frac{X(B) - X(A)}{S_E^C(A) - S_E^C(B)} < \alpha$$

Optimization considerations may lead the authorities to request an ambition level which is believed to be reasonably achievable because it is technologically and commercially available on the above conditions. Rather than requesting that particular technological solution, however, which may be an unduly restrictive requirement, the authorities may prefer to express the request in terms of authorized release limits that correspond to the optimum value of S_E^C. These limits, derived from optimization considerations, would not have to be rigidly enforced in situations where the original conditions no longer apply and the enforcement would seem unreasonable on cost-effectiveness grounds. They are therefore not true "limits" in the same sense as the ICRP dose limits.

In Sweden, one example of such release limitation is the "norm release" of radioactive materials from nuclear power stations, as specified in the present Swedish regulation (Swedish National Institute of Radiation Protection 1977). This is a release limitation which is considered reasonably achievable. It corresponds to an annual effective dose equivalent commitment of 10 mrem to individuals in a postulated, hypothetical critical group under average meteorological conditions. It is not applied as a limit but as a reference level for reporting trends in the release situation to the authorities within time limits specified in the regulations. The authorities will decide upon the acceptability of continued operation if releases exceed the norm release, on the basis of cost-effectiveness considerations under the actual meteorological conditions, subject to the overriding condition of individual dose limitation derived from the ICRP dose limits.

The question again arises whether dose contributions in the distant future should be discounted in the assessment of the collective effective dose equivalent commitment S_E^C for optimization purposes. Should the value of α remain constant irrespective of the time of dose delivery?

Although this is also essentially a political rather than a scientific question, the scientific information might indicate the answer. For nuclides such as tritium, krypton-85, strontium-90 and caesium-137, the mean life seems short enough to call for an assessment of the complete dose commitment, without discounting any dose contributions. Carbon-14, in spite of its half-life of 5700 years, delivers a significant fraction of its complete dose commitment before the nuclide becomes less available through mixing in the oceans. It may therefore be included in the same category. The real question arises with such long-lived nuclides as iodine-129 (and, for that matter, with chemically toxic, non-radioactive substances with infinite "life").

In those cases, it must be realized that protection through retention will only affect a minor fraction of the complete dose commitment. Irrespective of whether it is retained or not, iodine-129 will be free in the environment after a million years and since its mean life is near 25 million years, it will still always be able to deliver almost its full potential dose commitment. That dose is therefore independent of protective measures and should not be included in the assessment of S_E^c in optimization assessments. It follows that the integration for S_E^c should only be carried out over that period of time over which it is believed that iodine-129 can be retained if removed, i.e. a time period of the order of, at most 100,000 years.

It is usually not realized that the diminishing yield with improved retention makes it of interest to consider even very low-efficiency retention systems. A decontamination factor of 10 may, from the radiation protection point of view, be just as good as a decontamination factor of 10,000; in the first case 90 per cent of the collective dose is removed, in the second case 99.99 per cent. The difference in protective result is only 10 per cent while the difference in cost may be very large. It may be found that, although it is far from reasonable to install a high-efficiency retention system, it may well be worthwhile and give nearly the same result to install a low-efficiency system.

For the long-lived radionuclides, it may be the global or the regional rather than the local component of the collective dose that dominates. This may call for further conditions in the release limitation, in addition to those related to the doses in the critical group. These conditions may simply be that the long-lived radionuclides must be retained to a specified degree, e.g. 90 per cent or 99 per cent.

Individual Dose Limitation for Critical Groups

In 1956, ICRP widened its recommendations to cover not only radiation workers but also members of the public, because of the increasing use of radiation sources and the possible risk of harmful stochastic effects. Since the end of the 1950s the annual dose limit to the public has been 500 mrem (5 mSv).

Like the dose limit for workers, this limit for the public was first considered to be the highest *acceptable* dose. With the emphasis shifted to the optimization of protection, however, the dose limits have now obtained a different meaning and rather indicate the lower limit of *unacceptability*. No doses even below the dose limits, are acceptable unless protection is optimized and the practice justified.

If the benefit and the detriment were equally shared by the same individuals, there would be no need for individual dose limitation (the third basic requirement listed earlier), since the best interest of each individual would be served by the two first requirements of justification of the practice and optimization of protection. In practice, however, such equal distribution of benefit and detriment rarely occurs and it is therefore necessary, in addition, to guarantee the safety of each individual by the additional risk limitation by means of individual dose limits.

In its recent Publication 26 (1977a), ICRP refers to the new problems caused by the need of realistic dose assessments in optimization evaluations. If margins of safety are introduced in the assessment of the dose which would be expected from a certain release of radioactive substances into the environment, a bias is introduced in the optimization assessment and the outcome would not be correct. It is therefore expected that there will be a trend towards more realistic dose assessments. In those cases when some individuals might be exposed over many years, it would therefore be prudent to apply a more restrictive dose limitation, and ICRP recommends an annual dose of 1 mSv (100 mrem) for this purpose. This dose limit would be relevant in most cases of operation of polluting installations, when authorized release limits are derived.

The ICRP dose limit is intended to limit the total exposure from all controllable sources of radiation with the exception of medical exposures of patients. Patients are exceptional because they usually have both the benefit and the harm of the exposure and dose limits would therefore not be meaningful. Since many sources may expose one and the same individual, the operative authorized limit for each source must be less than the overall dose limit. In the Nordic countries (Nordic Radiation Protection Institutes 1976), and in the Swedish regulations (Swedish National Institute of Radiation Protection, 1977), an annual limit of the effective dose equivalent at 0.5 mSv (50 mrem) is indicated as appropriate for nuclear power stations (with the reservation that the risk-benefit balance of interventions must always be checked).

From such an operative dose limit, a corresponding release limit can be derived. The procedure of applying the release limit in practice, with a number of different radionuclides, has been described in an IAEA expert group report (1978), and in explanations to the Swedish release regulations. With some simplification the release limit R_L would be the combination of releases (R_i) of the various radionuclides i. These may be released from different release points j (into air or water) and cause exposures of various population groups k. In principle, dose factors f_{ijk} may be derived which give the effective dose equivalent commitment for members of group k from a unit release of radionuclide i from point j. Each group k would require a release limitation $R_{Lk} = (R_{ij})_{Lk}$ derived from:

$$\sum_i \sum_j f_{ijk} R_{ij} = H_{EL}$$

where H_{EL} is the authorized dose equivalent limit (Lindell, 1977). This condition must be satisfied for all groups k. Since it may be difficult to identify all groups k which might be limiting under various release conditions, it may be reasonable to simplify the calculation by using only the factors f_{ij} which represent the group k_{ij} that gives the highest doses per unit release of nuclide i from point j. The resulting release limit is overrestrictive, since the summation may add dose contributions to different groups of people for each nuclide. This conservatism, however, might be acceptable because of the substantial simplification.

The question arises over which period of time the dose commitment should be assessed. Since the whole purpose of the assessment is to find means to control the future dose rate, so that the accumulation of long-lived radionuclides will not result in the authorized dose limit being exceeded in the future, it is sufficient to assess the *incomplete* effective dose equivalent commitment, using an integration period which is assumed to be the total period of the practice. For this purpose various groups, e.g. UNSCEAR (1977), and the Nordic radiation protection institutes (1976), have assumed a total practice period of 500 years for nuclear power. This is unrealistically long, considering the amount of fuel available, but therefore also on the safe side in the assessment. If the dose commitment rather than the collective dose commitment is calculated, no assumptions have to be made on the future population size (but still on the population distribution). It can be shown that the maximum future annual effective dose equivalent in a specified population group is equal to the incomplete effective dose equivalent commitment to that group from one year of practice. This incomplete dose commitment is the time integral (over the period of practice, say 500 years) of the average effective dose equivalent rate in the group after one year of operation (UNSCEAR, 1977; Lindell, 1977; 1979). It can be shown that of the long-lived radionuclides released from the nuclear power industry, only carbon-14 would be expected to cause future dose rates in excess of the current authorized dose limits, if not retained. The justification for retention of the other long-lived nuclides would follow from optimization considerations rather than being imposed by the dose limits.

Control of Future Per Caput Doses in Large Populations

Carbon-14 is one example of a case where the limiting population group is not a local group but the whole world population. In this case, each individual will be exposed not just to one, but to many sources. There is a need to consider the overall effect of these sources as well as future developments which may increase the number of sources. This can be done on the basis of an assessment of the incomplete collective effective dose equivalent commitment per unit of practice.

If W denotes the global practice rate for the total world population of N individuals, the maximum future global *per caput* effective dose equivalent rate will be:

$$\bar{\dot{H}}_E \text{ max} = \dot{W} H_{E1}^{500} = \dot{W} \int_0^{500} \bar{\dot{H}}_{E1}(t)\, dt = \frac{\dot{W}}{N} S_{E1}^{500}$$

where H_{E1}^{500} is the incomplete (500 years integration) global effective dose equivalent commitment per unit practice, S_{E1}^{500} is the incomplete global collective effective dose equivalent commitment per unit practice, and $H_{E1}(t)$ is the global *per caput* effective dose equivalent rate caused by a unit of practice. For S_{E1}^{500}, the radiation protection institutes in the Nordic countries (1976) have recommended a goal of 1 manrem (10 millimansievert) per year of operation and for each installed MW of nuclear electric-power, for the whole fuel cycle. It follows that, at that level of achievement, a global nuclear electric power installation of 1 kw *per caput* would, after sufficient operation time to reach steady state conditions, cause a maximum global *per caput* effective dose equivalent rate of 10 microsievert (1 mrem) per year.

The expected contributions to S_{E1}^{500} from the release of long-lived radionuclides from the nuclear power industry, mainly from reprocessing plants, can be derived from data from the UNSCEAR reports

TABLE 1

Nuclide:	S_{E1}^{500} (manrem per MWyear electric energy)		
	no retention	90% retention of carbon-14	90% retention of all nuclides
tritium	0.1	0.1	0.01
carbon-14	1.2	0.12	0.12
krypton-85	0.1	0.1	0.01
iodine-129	0.01	0.01	0.001
all:	1.4	0.33	0.14

(1977). For light-water reactors, the following values would be expected (expressed per unit of produced electric energy rather than per year and installed power as in the above discussion; the difference is due to the fact that one year's operation of a 1000 MW reactor will not produce 1000 MW years of electric energy):

It seems therefore that there is no need of additional release limitation in order to respect the dose limits, provided that carbon-14 is retained to a degree of at least 90 per cent. This is also likely to be a requirement which would follow from the optimization assessment of retention systems.

References

Beninson, D. (1978a). Collective dose commitment. Working document for the OECD/NEA "Effluent Study" (in press).

Beninson; D. (1978b). Limitation of the release of radioactive effluents. Working document for the OECD/NEA "Effluent Study" (in press).

International Atomic Energy Agency (1977). "Optimization of radiation protection". Report from an expert panel in Tehran, (in press).

International Atomic Energy Agency. (1978). Principles for Establishing Limits for the Release of Radioactive Materials into the Environment. Report from an expert panel in Vienna, IAEA Safety Series No. 45.

International Commission on Radiation Units and Measurements. (1971). "Radiation Quantities and Units". ICRU Report 19. Washington, D.C.

International Commission on Radiation Units and Measurements. (1976). "Conceptual Basis for the Determination of Dose Equivalent". ICRU Report 25. Washington DC.

International Commission on Radiological Protection (1955). Recommendations of the International Commission on Radiological Protection. *British Journal of Radiology* Supplement No. 6.

International Commission on Radiological Protection (1958). "Recommendations of the International Commission on Radiological Protection" (report). Pergamon Press, London.

International Commission on Radiological Protection (1966). ICRP Publication No. 9 (report). Pergamon Press, Oxford.

International Commission on Radiological Protection. (1973). "Implications of the Commission's Recommendations that Doses be Kept as Low as Readily Achievable: ICRP Publication 22" (report). Pergamon Press, Oxford.

International Commission on Radiological Protection (1977a). "Recommendations of the International Commission on Radiological Protection: ICRP Publication 26" (report). Pergamon Press, Oxford.

International Commission on Radiological Protection (1977b).

"Problems involved in developing an index of harm: ICRP Publication 27" (report). *In* "Annals of the ICRP". Pergamon Press, Oxford.

International Commission on Radiological Protection (1978). "Statement from the 1978 Stockholm meeting of the "ICRP". (report). "Annals of the ICRP", Vol. 2, No. 1. Pergamon Press, Oxford.

Jacobi, W. (1974). How shall we combine the doses to different body organs? (conference paper). *In* "Selected Papers of the International Symposium of Aviemore". Commission of the European Communities.

Lindell, Bo (1977). New approaches to deriving limits of the release of radioactive material into the environment. *In* "Proceedings of the IAEA International Conference on Nuclear Power and its Fuel Cycle". Salzburg, Austria.

Lindell, Bo (1978a). New trends in radiation protection (conference paper). *In* "Proceedings of an International Symposium on Current Topics in Radiobiology and Photobiology". Academia Brasileira de Ciencias, Rio de Janeiro, Brazil.

Lindell, Bo. (1979). Source-related detriment and the commitment concept: applying the principles of radiation protection to non-radioactive pollutants. *Ambio 7*, 250-259.

Nordic Radiation Protection Institutes (1976. "Report on the Applicability of International Radiation Protection Recommendations in the Nordic Countries". Published in book form by the radiation protection institutes in Denmark, Finland, Iceland, Norway and Sweden.

Swedish Energy Commission (1978). "Radioaktiva ämnens hälso- och miljöeffekter" (report in Swedish, summary in English). Background material for the expert group on safety and environment, Swedish Department of Industry.

Swedish National Institute of Radiation Protection (1977). "Limitation of Releases of Radioactive Substances from Nuclear Power Stations" (regulation). Swedish regulations with an introduction on their background and purposes, English translation, Stockholm.

United Nations Scientific Committee on the Effects of Atomic Radiation (1977). "Sources and Effects of Ionizing Radiation". (report). UNSCEAR report to the U.N. General Assembly, New York.

U.S. Nuclear Regulatory Commission (1976). Title 10-chapter 1, Code of Federal Regulations, part 50, Appendix 1.

ON THE ASSESSMENT OF GENETIC AND CARCINOGENIC EFFECTS

Particularly with Respect to Chemicals Associated with Combustion Emissions in the Oxidation Fuel Cycles

Lars Ehrenberg and Göran Löfroth

*Department of Radiobiology,
University of Stockholm, Sweden.*

Energy production may give rise to physical and chemical factors which can cause impairment of the health in exposed populations. Outstanding examples are nuclear energy, through ionizing radiation, and oxidation energy, through a number of chemical compounds some already known and others, still unknown.

Combustion Emissions

Combustion emissions consist of both particulate matter and gases. The amount and composition of the emissions vary with the fuel, firing conditions and purification methods used for the flue-gas.

Particulate matter may range in diameter from several hundred µm to much less than one µm, and is composed of mineral constituents originating from the fuel, soot (often in the form of graphite), and adsorbed organic compounds including polycyclic organic matter (POM). POM includes polycyclic aromatic hydrocarbons (PAH). POM is to a large extent associated with particles less than 10 µm, including respirable particles, i.e. those less than 2-3 µm.

Gaseous components include carbon oxides, nitrogen oxides, sulphur oxides (if sulphur is present in the fuel), low molecular weight hydrocarbons (e.g. methane, ethene, propene etc.) and a variety of their oxidation products, aromatic hydrocarbons (e.g. benzene) and POM. Some metal and metalloid compounds may also appear in the gas phase.

Following emission into ambient air the pollutants undergo physical and chemical changes. Particles may aggregate to larger particles and compounds in the gas phase may adsorb to particles. The particles can – depending on size – be airborne for a considerable time and be transported over long distances before they are deposited on various surfaces. The oxides CO, NO and SO_2 are oxidized and so are also organic compounds. Particularly noteworthy is the photochemical production of oxidants involving the three-component system of hydrocarbons, nitrogen oxides and sunlight which give rise to a number of intermediate compounds.

Emission controls have in the past largely focused on gross particulate matter and SO_2 for stationary power plants and, in addition, on CO, NO_x and vapour-phase light hydrocarbons for motor vehicles.

Mutagenic and Carcinogenic Components

It should be emphasized that in order to assess a particular health hazard the responsible agents should be defined. Regarding genetic and carcinogenic, i.e. genotoxic, hazards due to combustion emissions it is at present neither possible to say that we are aware of all responsible agents or classes of compounds nor possible to say that any of the known classes of compounds are major contributors to the total hazard. From this follows the simple fact that a given decrease of the emission of certain components (e.g. particulate matter or PAH) may not necessarily mean that the total genotoxic hazard has decreased by the same factor or even by a significant amount.

Some known and potential genotoxic components associated with combustion emissions are given in Table 1. Unknown components which may include e.g. aldehydes,unsaturated oxo-compounds, ozone, peroxides, organic nitrates and nitrites etc., should be added to the table when recognized.

Dose Response Relationships

The effects caused by physical and chemical factors are largely an all or none effect for the individual molecule, individual cell or individual person. On a population basis, however, the effects are generally graded with respect to the received dose. The dose-response relationships have different forms for different end-points.

Non-genotoxic Effects

The classical toxicological dose-response curve is shown in Figure 1. It includes a threshold below which no effect is detected (present). This dose-response relationship is applicable to a number of components in the oxidation fuel cycles, such as carbon monoxide, certain systemically acting metal-compounds and probably also many of the irritants.

We are, however, not living in a perfect society, and large populations may be exposed to levels above the threshold for some components. This means that the dose-response system shifts as in Figure 2, and there is a detectable response to an infinitely small increase in the dose of these components.

Genotoxic Effects

According to current scientific knowledge, genotoxic effects,

TABLE I

Some Known and Potential Genotoxic Components Associated With Combustion Emissions

Component class	Example	Present as: P and/or G*	Mode of action
Alkenes	Ethene, Propene	G	Metabolism to epoxides which are alkylating compounds
Halogenated hydrocarbons	Methyl chloride	G	Unknown, alkylating or metabolism to alkylating compounds
Aromatic hydrocarbons	Benzene	G	Probably metabolism to epoxides
Polycyclic aromatic hydrocarbons (PAH)	Benz (a) anthracene, Cyclopenta (c,d) pyrene, Dibenzo (a,h) anthracene, Benzo (c) acridine	P/G	Metabolism to reactive inter-mediates, including epoxides
Non-PAH polycyclic organic matter (POM)	Not identified	P/G	Unknown

contd overleaf

TABLE 1 (Contd)

Component class	Example	Present as: P and/or G*	Mode of action
Metal compounds	As, Be, Cd, Cr, Ni (no speciation)	P/G	Unknown; some may have bio-chemical effects causing error in DNA replication
Nitrosamines	Dimethylnitrosamine	n.d. §	Metabolism to reactive species
Alkyl nitrites	Methyl nitrite	n.d.	Formation of nitroso compounds? transport of nitrite through membranes.

Unknown – see text

*P = particulate matter, G = gas phase
§ n.d. = not detected, which in these cases largely is the same as not determined.

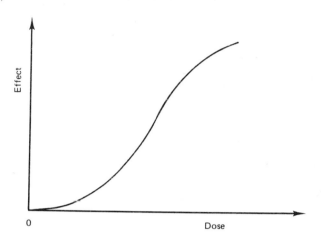

Fig. 1 The classical toxicological dose-response curve.

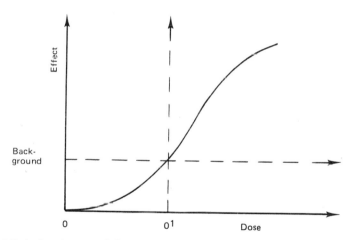

Fig. 2 High background levels can shift the classical dose-response curve to where small increases in dose elicit a response.

mutations in gonads, in embryonic cells and in somatic cells are linearly dependent on the dose at low doses. This kind of dose-response curve is shown in Figure 3.

This situation applies to substances which themselves (or their products) are directly capable of modifying the genetic code. It does not apply to factors which only cause changes in enzyme levels by inhibition or induction, as such effects probably have a threshold from the cell's point of view.

The use of linearity at low doses is thus not just an operational device for arriving at "safe levels" by overestimating the effects of a

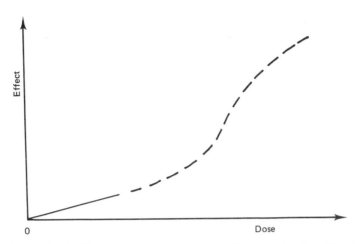

Fig. 3 Probable dose-response curve for genotoxic effects.

low dose.

It would appear that induction of cancer comprises several steps, including initiation and promotion. If initiation is limiting in a given situation, the dose response curve has a linear component at low doses. Some agents are carcinogenic through promotion, which may have several mechanisms, with thresholded dose response curves (Figure 1). In the situation of man, with a permanent exposure to initiating and promoting events, an addition of a new promoting agent may give a response starting from 0' in Figure 2, i.e. without any threshold (Crump et al., 1976).

Ionizing radiation is the only factor so far which has been evaluated with respect to stochastic effects with any claim to accuracy.

Assessment Methods

There are a number of methods for the monitoring and quantitative assessment of genotoxic effects. These methods are listed in Table 2.

It is worth noting that risk coefficients for *radiation induced* cancer have been evaluated from human epidemiological data, with little and doubtful contributions from animal experiments. Risks for radiation-induced heritable damage, on the other hand, are based on animal experiments, simply because epidemiological data do not exist (possibly except certain calculations of the lower limit of the doubling dose in Japanese A-bomb survivors). This leads to an enormous uncertainty, especially, with regard to consequences of mutation in the human germ-line.

Chemical carcinogenesis in man is concluded qualitatively from epidemiological studies in certain exposure situations, but, because of lack of knowledge of exposure doses in the past, risk coefficients can hardly be suggested. The difficulties of translating risks from animal experiments to man (cf. Ehrenberg and Holmberg, 1978) make experimental risk coefficients very uncertain. The few existing epidemiological studies of offspring from exposed fathers concern mostly foetal losses; properties of the compounds (e.g. vinyl chloride) make it uncertain that a mutational mechanism has been causative.

It has been indicated that the genetic risk from a unit dose of a chemical can be expressed by a general concept of radiation-risk-equivalent ("rad-equivalent" or "Sievert-equivalent"). In the range of doses where the responses can be considered linear (Figures 2 and 3) this gives a procedure to estimate chemical risks by using the radiation-risk coefficients of UNSCEAR[1] etc. In the problem of comparing nuclear and oxidation energy this procedure has the specific advantage of making the relative risks independent of errors in the risk coefficients and of uncertainties about consequences of heritable damage. It must be stressed, however, that considerable work remains before this procedure can be applied generally; correction factors for distribution of the dose in tissue, for steric effects, etc. have to be evaluated.

Concluding Remarks

It becomes evident that much is unknown about genotoxic compounds associated with the oxidation fuel cycles. Although much importance has been put on PAH, and particularly benzo(a)pyrene, this class of compounds is not necessarily a major factor. A much broader approach must be made including search for yet unidentified genotoxic compounds.

Exposure by inhalation may not necessarily be the only contributory route as at least two classes of relatively stable compounds (PAH and metals) are known to be deposited on surfaces including food crops. Deposition of relatively stable compounds (again e.g. PAH and metals) in sediments and soil may also involve genotoxic hazards for future generations.

Studies are needed to quantify emission factors and concentrations in the environment as well as the stability and transportation of emitted compounds and their reaction products in order to assess population doses, especially collective dose-commitments.

Methods for the quantitative assessment of genotoxic risks must be further developed. This concerns especially analytical methods.

Within ten to twenty years we will no doubt be able to determine the reaction products with blood proteins of most of the chemicals that constitute a genotoxic risk today. We think it is imperative that

---

[1] U.N. Scientific Committee on the Effects of Atomic Radiation.

TABLE 2

Methods for the Assessment of Genotoxic Effects

Method	End-point	Some major difficulties	Advantages	Some references
Epidemiology	Cancer	Identification of proper control groups; difficulties of assessing exposure doses, partly because effects may be due to conditions 10–50 years before detection.	Gives an independent control of risk estimates by other methods.	Hueper et al. (1962) Task group Report (1978)
"	Chromosomal aberrations *in vivo*	Relationship to genetic risks variable and integral doses back in time.	For some compounds may be used to assess in time.	
"	Gonadal mutations	(Has begun with a few compounds)		Infante et al. (1976).
Animal bioassays	Several	Translation to man; some tests are relatively expensive.	Qualitative detection of genotoxic compounds; detection of promotors.	

TABLE 2 (Contd)

Method	End-point	Some major difficulties	Advantages	Some references
Short-term bioassays	Several	Translation to man.	Qualitative detection of genotoxic compound; 'quantitative' comparison between samples of similar composition; relatively inexpensive.	Brusick, (1978); Harnden, (1978); Löfroth, (1978).
Tissue dose determination	Adduct formation with nucleophilic centres	At present limited to certain organic compounds and their metabolites.	May be utilized to determine doses directly in man; collective doses and risks may be evaluated; the method is, in principle, very sensitive	Calleman et al. (1978) Ehrenberg et al. (1977); Segerbäck et al. (1978).

we give a basis to future epidemiologists by creating a repository of frozen blood samples that characterizes today's environment. We would end by suggesting a project to develop this idea further.

Acknowledgement

Work leading to this review has been supported by the Swedish Board of Occupational Safety and Health, Swedish Work Environmental Fund, National Swedish Environment Protection Board, Swedish Natural Science Research Council and the (former) Swedish Atomic Research Council.

References

Brusick, D.J. (1978).   The role of short-term testing in carcinogen detection. *Chemosphere* **7**, 403-417.

Calleman, C.J., Ehrenberg, L., Jansson, B., Österman-Golkar, S., Segerbäck, D., Svensson, K. and Wachtmeister, C.A. (1978). Monitoring and risk assessment by means of alkyl groups in hemoglobin in persons occupationally exposed to ethylene oxide.   *Journal of Environmental Pathology and Toxicology* **2**, 427-442.

Crump, K.S., Hoel, D.G., Langley, C.H. and Peto, R. (1976). Fundamental carcinogenic processes and their implications for low dose risk assessment. *Cancer Research* **36**, 2973-2979.

Ehrenberg, L. and Holmberg, B. (1978). Extrapolation of carcinogenic risk from animal experiments to man.   *Environmental Health Perspectives* **22**, 33-35.

Ehrenberg, L., Österman-Golkar, S., Segerback, D., Svensson, K. and Calleman, C.J. (1977).   Evaluation of genetic risks of alkylating agents. III. Alkylation of hemoglobin after metabolic conversion of ethene to ethene oxide.   *Mutation Research* **45**, 175-184.

Harnden, D.G. (1978).   Relevance of short-term carcinogenicity tests to the study of the carcinogenic potential of urban air. *Environmental Health Perspectives* **22**, 67-70.

Hueper, W.C., Kotin, P., Tabor, E.C., Payne, W.W., Falk, H. and Sawicki, E. (1962).   Carcinogenic bioassays on air pollutants. *Archives of Pathology* **74**, 89-116.

Infante, P.F., McMichael, A.J., Wagoner, J.K., Waxweiler, R.J. and Falk, H. (1976).   Genetic risks of vinyl chloride.   *The Lancet* **1**, 734-735.

Löfroth, G. (1978).   Mutagenicity assay of combustion emissions. *Chemosphere* **7**, 791-798.

Segerbäck D. *et al.* (1978).   Evaluation of genetic risk of alkylating agents.   IV.   Quantitative determination of alkylated amino acids in hemoglobin as a measure of the dose after treament of mice with methyl methanesulfonate.   *Mutation Research* **49**, 71-82.

Task Group on Air Pollution and Cancer (1978). Risk assessment methodology and epidemiological evidence. *Environmental Health Perspectives* **22,** 1-10.

IMAGES OF DISASTER: PERCEPTION AND ACCEPTANCE OF RISKS FROM NUCLEAR POWER

Paul Slovic, Sarah Lichtenstein and Baruch Fischhoff

*Decision Research, A Branch of Perceptronics,
Eugene, Oregon, U.S.A.*

"I do not know if our occupation (of the Trojan Nuclear Plant) will be effective in educating the people to the dangers of nuclear radiation. I hope so. I was terribly uncomfortable in jail for five days in a six-person cell with 40 other women, three towels for all of us, 16 mattresses, 16 blankets, and no air.

I could go on and on . . . but to what use? Perhaps only that you might take notice that we give up our loving homes and gardens and children to be arrested, jailed, fined, brutalized, and degraded. WHY? WHY? Because we truly believe that the planet and the human race is in grave danger because of nuclear power plants. Will you hear us?"   (Published letter to the Governor of Oregon from a nuclear power protester – 1978)

Our concern in this paper is with public response to the risks of nuclear energy. The topic is a vital one. Writing recently in the *American Scientist,* Alvin Weinberg (1976) observed:

"As I compare the issues we perceived during the infancy of nuclear energy with those that have emerged during its maturity, the public perception and acceptance of nuclear energy appears to be the question that we missed rather badly . . . This issue has emerged as the most critical question concerning the future of nuclear energy."

At present, the nuclear industry is foundering on the shoals of adverse public opinion. A sizeable and tenacious opposition movement has been responsible for costly delays in the licensing and construction of new power plants in the United States and for political turmoil in several European nations. Any attempt to plan the future role of nuclear power must consider the determinants and possible course of this opposition.

Although opposition to nuclear power has many causes, concerns about safety undoubtedly play a major role (see, e.g., Bronfman and Mattingly, 1976;   Hohenemser, Kasperson and Kates, 1977; Melber, Nealey, Hammerslea and Rankin, 1977;   Otway, Maurer and Thomas, 1978).   Over the past several years, we and others have been systematically investigating public perceptions of risk from nuclear power and other hazardous activities.   From this research, we have pieced together a quantitative description of the attitudes, perceptions and expectations of some members of the anti-nuclear public. The images of potential nuclear disasters that have been formed in these people's minds are remarkably different from the assessments put forth by most technical experts. In this paper, we shall describe these images and speculate on their origins, permanence, and implications.

Many of the studies described below have *not* examined a representative sample of the public at large, or even of the anti-nuclear public.   Instead, much of the new data comes from two populations, both known to be strongly anti-nuclear by their voting behaviour in the 1976 nuclear power referendum in Oregon.   One population consists mainly of students from the University of Oregon.   The second consists of members of the Eugene, Oregon, League of Women Voters. Members of the League (hereafter designated as the LOWV) were not selected because of their generally anti-nuclear views, but because they constitute a thoughtful, articulate, and influential group of private citizens.   In general, we have found many points of similarity between the views expressed by these particular individuals and those expressed by more representative samples of people in the United States and abroad.

Perceived Risks and Benefits

How is nuclear power viewed, relative to other activities?   One answer to this question is provided in a study by Fischhoff, Slovic, Lichtenstein, Read and Combs (1978) in which 40 LOWV members judged the risk of dying (across all U.S. society as a whole) as a consequence of nuclear power and 29 other activities and technologies. Each activity appeared on a 3 x 5 inch card.   Respondents first studied the items individually, thinking of all the possible ways someone might die from each.   Then they ordered the items from least to most risky and assigned numerical risk values by giving a rating of 10 to the least risky item and making the other ratings accordingly.   For example, they were told "a rating of 12 indicates that the item is 1.2 times as risky as the least risky item (i.e., 20 per cent more risky).   A rating of 200 means that the item is 20 times as risky as the least risky item, to which you assigned a 10 . . ."   They were urged to cross-check and adjust their numbers until they believed they were right.

After rating the risks from the various activities, the respondents indicated whether each risk was: (a) presently acceptable, (b) low enough so that it could even be greater before serious action (such as legislation) would need to be taken to control it, or (c) too high so that such action needed to be taken now. If categories (b) or (c) were selected, the respondent was asked to indicate how many times riskier or safer that activity's risk level needed to be to reach an "acceptable level."

An additional group of 36 LOWV members rated the present *benefits* to society from each activity and technology by means of the same scaling technique that was used to determine perceived risk. They also rated the acceptability of current risk levels.

The results from this study, shown in Table 1, give some clues as to why 95 per cent of these individuals voted to curtail development of nuclear power in the state-wide referendum held several months after these data were collected. For one, the benefits of nuclear power appeared unappreciated, being lower than those of home appliances, bicycles, and general aviation. Perhaps this is because nuclear power was seen merely as a supplement to other sources of energy, which themselves were viewed as adequate. Nuclear power's low perceived benefit in these studies is consistent with more general conclusions from a representative survey of the adult U.S. population by Pokorny (1977). This survey showed that people do not see nuclear power as a vital link in the supply of basic energy needs. A second major result was that risks from nuclear power were judged to be extremely high. Only motor vehicles, which take about 50,000 lives each year, were viewed as comparably risky. Third, participants in this study wanted nuclear power to be far safer than they perceived it to be at the present time (29 times safer to be exact). When this study was repeated with a sample of 69 students, the results (not shown here) were quite similar to those from the LOWV sample.

Risk Characteristics

The same LOWV members who judged risks and benefits in the study by Fischhoff *et al.* (1978) also rated nuclear power and the 29 other hazardous activities on nine characteristics which have been hypothesized to influence perceptions of actual or acceptable risk (e.g. Lowrance, 1976). These rating scales are described in Table 2.

The "risk profiles" derived from these ratings showed that nuclear power had the dubious distinction of scoring at or near the extreme high-risk end for most of the characteristics. Its risks were seen as involuntary, unknown to those exposed or to science, uncontrollable, unfamiliar, potentially catastrophic, severe (likely to be fatal rather than injurious), and dreaded. Its spectacular and unique risk profile is contrasted in Figures 1 and 2 with non-nuclear electric power and another radiation

TABLE 1

Perceived Risk and Need for Risk Adjustment for 30 Activities and Technologies*

Activity or Technology	Perceived Benefit	Perceived Risk	Need for Risk Adjustment†
1. Alcoholic beverages	41	161	4.4
2. Bicycles	82	65	1.5
3. Commercial aviation	130	52	1.3
4. Contraceptives	113	50	2.0
5. Electric power	274	52	1.0
6. Fire fighting	178	92	1.1
7. Food colouring	16	31	3.0
8. Food preservatives	44	36	2.7
9. General aviation	53	114	2.1
10. Handguns	14	220	17.3
11. H.S. & college football	35	37	1.7
12. Home appliances	133	25	1.1
13. Hunting	30	82	2.5
14. Large construction	142	91	1.7
15. Motorcycles	29	176	5.3
16. Motor vehicles	187	247	6.1
17. Mountain climbing	28	68	1.0
18. Nuclear power	52	250	29.0
19. Pesticides	87	105	9.5
20. Power mowers	30	29	1.5
21. Police work	178	111	1.8
22. Prescription antibiotics	209	30	1.3
23. Railroads	185	37	1.2
24. Skiing	38	45	1.0
25. Smoking	20	189	15.2
26. Spray cans	17	73	7.8
27. Surgery	164	104	1.9
28. Swimming	68	52	1.0
29. Vaccinations	194	17	.8
30. X-rays	156	45	1.7

*Data adapted from Fischhoff et al. (1978).
†Values of 1.0 indicate that the activity is presently at an acceptable level of risk. Values greater than 1.0 mean the activity needs to be safer by the factor indicated in the column; values less than 1.0 mean the activity could be riskier and still be acceptable to society.

TABLE 2

Risk Characteristics Rated by LOWV Members and Students

Voluntariness of risk

Do people face this risk voluntarily? If some of the risks are voluntarily undertaken and some are not, mark an appropriate spot towards the center of the scale.

| risk assumed voluntarily | 1 | 2 | 3 | 4 | 5 | 6 | 7 | risk assumed involuntarily |

Immediacy of effect

To what extent is the risk of death immediate — or is death likely to occur at some later time?

| effect immediate | 1 | 2 | 3 | 4 | 5 | 6 | 7 | effect delayed |

Knowledge about risk

To what extent are the risks known precisely by the persons who are exposed to those risks?

| risk level known precisely | 1 | 2 | 3 | 4 | 5 | 6 | 7 | risk level not known |

To what extent are the risks known to science?

| risk level known precisely | 1 | 2 | 3 | 4 | 5 | 6 | 7 | risk level not known |

Control over risk

If you are exposed to the risk, to what extent can you, by personal skill or diligence, avoid death?

| personal risk can't be controlled | 1 | 2 | 3 | 4 | 5 | 6 | 7 | personal risk can be controlled |

Newness

Is this risk new and novel or old and familiar?

| new | 1 | 2 | 3 | 4 | 5 | 6 | 7 | old |

Chronic-catastrophic

Is this a risk that kills people one at a time (chronic risk) or a risk that kills large numbers of people at once (catastrophic risk)?

| chronic | 1 | 2 | 3 | 4 | 5 | 6 | 7 | catastrophic |

Common-dread

Is this a risk that people have learned to live with and can think about reasonably calmly, or is it one that people have great dread for — on the level of a gut reaction?

| common | 1 | 2 | 3 | 4 | 5 | 6 | 7 | dread |

Severity of consequences

When the risk from the activity is realized in the form of a mishap or illness, how likely is it that the consequence will be fatal?

| certain not to be fatal | 1 | 2 | 3 | 4 | 5 | 6 | 7 | certain to be fatal |

technology, X-rays, both of whose overall risks were judged to be much lower. Both non-nuclear electric power and X rays were judged much more voluntary, less catastrophic, less dreaded, and more familiar than nuclear power. Nuclear power was rated far higher on the characteristic "dread" than any of the 29 other hazards studied. This may stem from the association of nuclear power with nuclear weapons and from fear of radiation's invisible, permanent bodily contamination that causes genetic damage and cancer (Lifton, 1976; Pahner, 1975). A replication of this study with the student sample produced results virtually identical to those of the LOWV members.

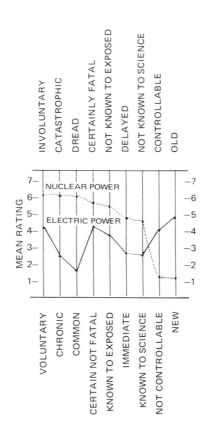

Fig. 1 Comparison between nuclear power and non-nuclear electric power on nine risk characteristics (from Fischhoff et al., 1978).

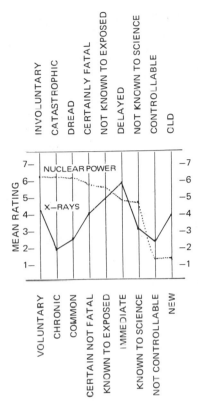

Fig. 2 Comparison between nuclear power and X-rays on nine risk characteristics (from Fischhoff *et al.*, 1978).

Why is Nuclear Power Thought So Dangerous?

Although questions of safety seem to be pre-eminent in the nuclear debate, it is important to recognize that opposition to nuclear power is an organized political movement fuelled by many other concerns besides safety (Bronfman and Mattingly, 1976; Otway, Maurer and Thomas, 1978; Wilkes, Lovington, Horne, Pulaski and Poole, 1978). While some nuclear opponents are motivated primarily by fears of routine or catastrophic radiation releases, others join the movement because they are disenchanted with growth, centralization, corporate dominance, technology, or government and its institutions. The latter individuals may argue questions of safety because they view the hazardousness of nuclear power as its "Achilles Heel." While the studies presented here are not directly concerned with this larger political context, they do highlight the special qualities of nuclear power that cause political opposition to be focused around considerations of risk.

Seriousness of a Death from Nuclear Power

One hypothesis we studied is that people think nuclear risks are great because they believe a death from nuclear power is somehow much worse than a death from other activities. For example, deaths from involuntary or dread risks might be given extra weight, causing people to view a death from nuclear power as particularly serious.

We tested this hypothesis by asking new groups of students and LOWV members to evaluate the seriousness of death from the 30 activities and technologies (plus four additional hazards) according to the following instructions:

"Decision makers often have a very difficult time determining whether or not the risk of death from some new product or technology is acceptable to society. One question that comes up repeatedly when people try to assess the acceptability of risk is whether or not all deaths are equally bad regardless of their cause. Some people say that a life is a life, no matter how it is lost. Other people disagree with this notion. They argue that deaths from some causes should receive a lesser weight in our societal decisions.

"Listed on the following pages are some specific causes of death. We'd like to know how you think society should weight these causes of death.

Assume there is some standard unit of loss for a single death. This unit might be represented by an amount of money, such as one million dollars lost. Call this unit a *standard loss*. How should this standard loss be altered for each of the circumstances we list?

Example: Death due to Cause A

Loss is less than standard Divide standard by ?	Standard loss is appropriate	Loss is greater than standard Multiply standard by ?

In considering the value of a death due to Cause A, you have 3 options:

1) If you believe that the standard loss is appropriate, simply put an X below the middle option. If you believe that all kinds of deaths are equally bad, you should always used this middle option.

2) If you believe that deaths due to Cause A are particularly bad for society so that the standard loss needs to be increased by 10%, 50%, a factor of 2, 10, etc., put the necessary multiplier in the *right hand* column. (For an increase of 10% enter 1.1; for an increase of 50% enter 1.5; for two- or ten-fold increases, enter 2 or 10, respectively, etc.).

3) If you believe that deaths due to Cause A are not as bad as most deaths for society and should be weighted less than the standard loss, put the factor by which the standard should be reduced in the *left hand* column (1.1, 1.5, 2, 10, etc.). The standard will then be *divided* by this factor."

The results of this study, shown in Table 3, surprised us. Although a death from nuclear power was seen as the most serious kind of death by the students and third most serious for the LOWV members, the differences in seriousness, across the 30 hazards, were not that large.   In essence, our respondents appeared to be saying that all modes of death are about equally bad.   Results obtained when the standard loss was changed to $10,000 also supported this conclusion.

If extreme aversion to nuclear power cannot be attributed to the subjective seriousness of deaths due to that technology, perhaps they might stem from horrific, but non-fatal, consequences, such as genetic defects and malformed babies.   While we did not test specific non-fatal consequences of nuclear power, we did ask our judges to rate the social costs of several very serious non-fatal afflictions.   These included severe mental retardation, congenital blindness, and a Thalidomide baby (seal-like flippers instead of arms and legs).   The instructions were similar to those used for rating the social cost of death.   The results indicated that individual cases of these afflictions were not viewed as significantly more serious than deaths due to nuclear power or any of the other activities.   In sum, it seems unlikely that the aversiveness of nuclear power reflects extreme weighting of radiation-induced death or disability.

Expected Number of Deaths

Using a method very different from that just described, Otway, Maurer and Thomas (1978) also found that aversion to nuclear power was not due to overweighting of the seriousness of its consequences.   Specifically, they found that various forms of nuclear-induced mortality, morbidity, and environmental and social damages were judged about equally bad by pro-nuclear and anti-nuclear individuals. Where nuclear proponents and opponents did differ, however, was in their expectations about how likely these aversive consequences were to occur or how many of them there might be.   This suggests that one major source of aversion to nuclear power may be the expectation that this technology will cause an exceedingly large number of fatalities.

Evidence supporting this hypothesis comes from a study in which we asked 39 students and 38 LOWV members to estimate the frequency of death to be expected from the 30 activities and technologies shown in Table 1.   Specifically, we asked: "How

TABLE 3

Geometric Mean Judgements of Social Cost of Death

		Social Cost in 10^6 dollars (Students; N=77)	Social Cost in 10^6 dollars (LOWV; N=45)
1.	Alcoholic beverages	.91	.77
2.	Bicycles	1.25	1.27
3.	Cancer	1.37	1.22
4.	Commercial aviation	1.41	1.27
5.	Construction	1.15	1.25
6.	Contraceptives	1.80	1.39
7.	Dam failure	1.47	1.73
8.	Electric power (non-nuclear)	1.38	1.25
9.	Fire fighting	1.49	1.45
10.	Food colouring	1.60	1.75
11.	Food preservatives	1.72	1.83
12.	Handguns	1.48	2.52
13.	H.S. & college football	1.35	1.41
14.	Home appliances	1.25	1.19
15.	Homicide	1.97	3.08
16.	Hunting	1.05	.94
17.	Motorcycles	1.10	.97
18.	Motor vehicles	1.25	1.25
19.	Mountain climbing	1.04	.97
20.	Nuclear power	2.26	2.41
21.	Pesticides	1.73	2.05
22.	Police work	1.39	1.64
23.	Power mowers	1.22	1.28
24.	Prescription antibiotics	1.52	1.29
25.	Private aviation	1.07	1.06
26.	Railroads	1.29	1.16
27.	Skiing	1.17	1.12
28.	Smoking	.96	.70
29.	Spray cans	1.59	1.61
30.	Suicide	.95	.94
31.	Surgery	1.34	1.21
32.	Swimming	1.09	1.11
33.	Vaccinations	1.61	1.42
34.	X-rays	1.59	1.52

many people are likely to die in the U.S.A. next year (if next year is an average year) as a consequence of these activities and technologies?" All sources of deaths from an activity were to be considered. For example, the instructions specified that fatalities from non-nuclear electricity should consider mining of coal and other energy production activities as well as electrocution; motor-vehicle fatalities were to include collisions with bicyclists and pedestrians, etc. As a guideline, the instructions indicated that the total number of deaths in the U.S. averages about 2,000,000 per year.

In addition to estimating fatalities for an average year, the students and LOWV members were asked to provide a multiplying factor indicating how many times more deaths than the average would occur if next year was "particularly disastrous" for the activity being considered. This second estimate allowed concerns about catastrophic potential to be expressed. Mean fatality estimates and multiplying factors are presented in Table 4. Examination of these data shows that the expected number of fatalities in an average year was *smaller* for nuclear power than for any other activity or technology. However, nuclear power was in a class by itself with respect to the multiplying factor. More than 40 per cent of the respondents had multiplying factors for nuclear power that were greater than 1,000. For each individual respondent, an estimate of the expected number of fatalities in a *disastrous* year was calculated by applying that person's disaster multiplier to his or her fatality estimate for an average year. When this was done for nuclear power, almost 40 per cent of the respondents had estimates greater than 10,000 fatalities and more than 25 per cent had estimates exceeding 100,000 fatalities.

Disaster Scenarios

These extreme multiplying factors suggest that many people expect nuclear power to lead to disasters of immense proportions. In an attempt to better understand the nature of these concerns, we asked a new group of 28 students to write scenarios for nuclear power and commercial aviation describing their image of the maximum credible disaster that might be produced by each activity during their lifetime. The instructions read:

"In this task you will be asked to describe, for nuclear power and commercial aviation, your image of the *maximum credible disaster*. This should not be the biggest disaster that you can dream up, but rather the biggest disaster you seriously think might occur *during your lifetime*.

Simply describe your image of the maximum credible mishap or disaster in a brief written paragraph. The paragraph should indicate the circumstances of the mishap, the number of injuries or fatalities that might occur, and the size of

TABLE 4

Fatality Estimates and Disaster Multipliers
for 30 Activities and Technologies

Activity or Technology	Geometric Mean Fatality Estimates Average Year		Geometric Mean Multiplier Disastrous Year	
	LOWV	Students	LOWV	Students
1. Alcoholic beverages	12,000	2,600	1.9	1.4
2. Bicycles	910	420	1.8	1.4
3. Commercial aviation	280	650	3.0	1.8
4. Contraceptives	180	120	2.1	1.4
5. Electric Power	660	500	1.9	2.4
6. Fire fighting	220	390	2.3	2.2
7. Food colouring	38	33	3.5	1.4
8. Food preservatives	61	63	3.9	1.7
9. General aviation	550	650	2.8	2.0
10. Handguns	3,000	1,900	2.6	2.0
11. H.S. & college football	39	40	1.9	1.4
12. Home appliances	200	240	1.6	1.3
13. Hunting	380	410	1.8	1.7
14. Large construction	400	370	2.1	1.4
15. Motorcycles	1,600	1,600	1.8	1.6
16. Motor vehicles	28,000	10,500	1.6	1.8
17. Mountain climbing	50	70	1.9	1.4
18. Nuclear power	20	27	107.1	87.6
19. Pesticides	140	84	9.3	2.4
20. Power mowers	40	33	1.6	1.3
21. Police work	460	390	2.1	1.9
22. Prescription antibiotics	160	290	2.3	1.6
23. Railroads	190	210	3.2	1.6
24. Skiing	55	72	1.9	1.6
25. Smoking	6,900	2,400	1.9	2.0
26. Spray cans	56	38	3.7	2.4
27. Surgery	2,500	900	1.5	1.6
28. Swimming	930	370	1.6	1.7
29. Vaccinations	65	52	2.1	1.6
30. X-rays	90	40	2.7	1.6

the geographic region affected."

Some examples of the nuclear-power scenarios these instructions elicited are the following:

Subject 2: "An unpredicted failure occurs in the cooling system of a reactor. The critical point is attained where an explosion destroys the plant and emits masses of radioactive material into the atmosphere. The populations of neighbouring communities are decimated; 5,000 die."

Subject 19: "A core meltdown occurs in the reactor of a nuclear power plant located a few miles from a major city. The backup systems fail and a lot of deadly radiation escapes. The winds blow it toward the metropolitan area which has a population of 2 million. The nuclear fallout kills 200,000 people within a 35-mile radius of the plant."

The distributions of fatalities associated with the various scenarios are shown in Table 5. The most common scenarios for commercial aviation were based on crashes between two jumbo jets (usually Boeing 747s). The two extreme aviation disasters in Table 5 involved crashes into heavily populated areas. Overall, these estimates of our subjects are not much higher than those made by experts. The Reactor Safety Study (U.S. Nuclear Regulatory Commission, 1975) reports data suggesting that the probability of an aviation accident involving 1,000 deaths (our subjects' median estimate) occurring at least once during a 60-year lifetime is about .02 or .03. Thus, such a response is not unreasonable in light of our instructions to describe "the biggest disaster you seriously think might occur during your lifetime." Only three of our 29 subjects estimated aviation fatalities beyond the range considered as at least remotely possible by experts.

In contrast, the fatality estimates associated with nuclear power would not be considered reasonable by most technical experts. These estimates tended to be several orders of magnitude greater than those provided in the Reactor Safety Study (U.S. Nuclear Regulatory Commission, 1975). According to that study, the maximum possible accident, coincident with the most unfavourable combination of weather and population density, would cause about 3,300 prompt fatalities and 45,000 latent cancer fatalities with a probability of 5×10^{-9} per reactor year. Assuming 100 reactors operating over a period of 60 years, the probability of one or more such disasters is only .00003. Yet, a sizeable percentage of our scenario writers appear to expect a disaster of greater severity within their lifetimes.

Facing the Facts: Can the Public Be Unscared?

These results document what many other observers have suspected: there is an immense gap between the opinions of most technical experts and the views of the public regarding the risks from nuclear

P. Slovic et al.

TABLE 5

Fatality Estimates Associated with "Maximum Credible Disasters"
from Commercial Aviation and Nuclear Power

Commercial Aviation		Nuclear Power*	
50		0	
100		3	
100		10	
100		48	
138		60	
200		60	
220		1,000	
300		2,000	
hundreds		3,000	
431		5,000	
450		10,000	
500		20,000	median
600		25,000	values
700		30,000	
900	median	50,000	
1,000	values	100,000+	
1,000		200,000	
1,500		200,000	
2,000		250,000	
2,000		1,000,000	
2,000		1,000,000	
2,000		2,000,000	
2,000		10,000,000	
3,000		20,000,000	
3,500			
5,000			
10,000			
200,000			
300,000			

Note : Each value represents the expectations of a different respondent.

*Five persons did not provide numerical estimates with their scenarios. Of these, three wrote scenarios postulating world-wide contamination and death.

power. The contribution of the present study is the evidence that public fears are not derived from a feeling that nuclear power is a particularly aversive mode of death, but from the expectation that an enormous number of deaths are likely to result as a consequence of that technology. Many people, like the letter writer quoted in the introduction to this paper, apparently believe that nuclear power threatens the survival of "the planet and the human race."

Kasper has commented at length earlier in this book on the consequences of the discrepancy between expert and lay judgements of nuclear risk. One consequence is that experts, fearing overreaction by the public, feel forced to overstate the precision of their estimates. A more important consequence is the confusion and distrust on the part of a public which believes the risks to be vastly greater than the experts' assessments indicate. The experts, in turn, question the rationality of the public and decry the emotionalism stymying technological progress. Bitter and occasionally violent confrontations result.

Recognition of the "perception gap" has naturally resulted in the belief that the public must be educated about the "real" risks from nuclear power. One public opinion analyst put the matter as follows:

> "The biggest problem hindering a sophisticated judgement on this question is basic lack of knowledge and facts. Within this current attitudinal milieu, scare stories, confusion, and irrationality often triumph. Only through careful education of facts and knowledge can the people know what the real choices are and can thereafter make the decision wisely" (Pokorny, 1977).

A Pessimistic View

Our own view is that educational attempts designed to reduce the "perception gap" are probably doomed to failure. This pessimistic conclusion is based on two key aspects of the problem, one technical and one psychological. The technical problem is that the disputed risks are so improbable that they are not amenable to precise empirical verification. The psychological problem is that people's perceptions are not irrational but are based on normal ways of thinking which, when applied to the special qualities of nuclear risks, are likely to thwart attempts to modify beliefs.

The Technical Reality The technical reality is that there are few "cut and dried facts" regarding the probabilities of serious reactor mishaps. The technology is so new and the probabilities in question are so small that the risk estimates cannot be based on empirical observation. Instead, risk assessments must be derived from complex mathematical models and structures such as the fault trees and event trees used in the Reactor Safety Study to estimate the probability of a loss-of-coolant accident (U.S. Nuclear Regulatory Commission, 1975). Despite an appearance of objectivity, such assessments are inherently subjective. Someone, relying on judgement, must determine the structure of the analysis, including the ways that failure might occur, their relative importance, and their logical interconnections.

The difficulties of applying fault-tree and event-tree analysis have led many critics to question the validity of these methods (e.g., Bryan, 1974; Fischhoff, 1977; Primack, 1975). One major concern is that important initiating events or pathways to failure may be omitted, causing risks to be underestimated. Another problem is the difficulty of taking proper account of "common-mode failures." To insure greater safety, many technological systems are built with a great deal of redundancy. Should one crucial part fail, there are others designed either to do the same job or to limit the resulting damage. Since the probability of each individual part failing is very small, the probability of all failing, thereby creating a major disaster, should be extremely small. This reasoning is valid only if the various components are independent – that is, if what causes one part to fail will not automatically cause the others to fail. "Common-mode failure" occurs when the independence assumption does not hold. As an example, the discovery that a set of pipes in several nuclear plants were all made from the same batch of defective steel (*Eugene Register Guard*, Oct. 13, 1974) suggests that the simultaneous failure of several such pipes was not inconceivable. The same fire that caused the core to overheat in a reactor at Brown's Ferry, Alabama also damaged the electrical system needed to shut the plant down. Developing models to assess such contingencies is a very difficult enterprise.

Questions about completeness, treatment of common-mode failures, and related statistical problems have caused a recent review of the Reactor Safety Study (U.S. Nuclear Regulatory Commission, 1978) to conclude that the error bounds on the probabilities calculated in that study are greatly underestimated. This review has led the Nuclear Regulatory Commission to caution that the absolute values of risk reported in the Reactor Safety Study should not be used uncritically either in the regulatory process or for public policy purposes.

Holdren's scepticism regarding the defensibility of assessments of rare catastrophes summarizes the technical problem concisely:

> ". . . the expert community is divided about the conceivable realism of probability estimates in the range of one in ten thousand to one in one billion per reactor year. I am among those who believe it to be impossible *in principle* to support numbers as small as these with convincing theoretical arguments (that is, in the absence

of operating experience in the range of 10,000 reactor-years or more), even ignoring the crucial possibility of malevolence. The reason I hold this view is straightforward: nuclear power systems are so complex that the probability the safety analysis contains serious errors ... is so big as to render meaningless the tiny computed probability of accident." Holdren, 1976).

The Psychological Reality Public fears of nuclear power should not be viewed as irrational. In part, these fears are fed by the realization that (a) the experts have been wrong in the past, as when they irradiated enlarged tonsils or permitted people to witness A-bomb tests at close range, and (b) the experts are still disputing the "facts" about risks from nuclear power. Furthermore, when examined in detail, people's fears appear to reflect fundamental ways of thinking that usually lead to satisfactory judgements and decisions. However, when applied to low-probability, high-consequence risks such as those encountered with nuclear power, these modes of thinking lead to heightened fears and reluctance to change beliefs in the face of new evidence.

One mode of thought that provides insight into perceived nuclear risks is a judgemental rule or strategy known as the *availability heuristic* (Tversky and Kahneman, 1974; Slovic, Fischhoff, Lichtenstein and Hohenemser, 1979). This rule leads people to judge an event as likely or frequent if instances of it are easy to imagine or recall. Frequently occurring events are generally easier to imagine and recall than rare events; thus, reliance on availability is typically an appropriate mental strategy. However, memorability and imaginability are also affected by numerous factors not related to likelihood. As a result, this natural way of thinking leads people to exaggerate the probabilities of events that are particularly recent, vivid, or emotionally salient. Certainly, the risks from nuclear power would seem to be a prime candidate for enhancement by the availability heuristic, because of the extensive media coverage they receive and their association with the vivid, imaginable dangers of nuclear war. As Zebroski (1976) noted, "fear sells"; the media dwell on potential catastrophes, not on the successful day-to-day operations of power plants.

One subtle and disturbing implication of the availability heuristic is that any discussion of low-probability hazards, regardless of its content, may increase the memorability and imaginability of those hazards and, hence, increase their perceived risks. This possibility poses a major barrier to open, objective discussions of nuclear safety. Consider an engineer demonstrating the safety of disposing nuclear wastes in a salt bed by using the fault tree shown in Figure 3 to point out the improbability of the various ways radioactivity could be released. Rather than reassuring the audience, the presentation might lead them to think: "I didn't

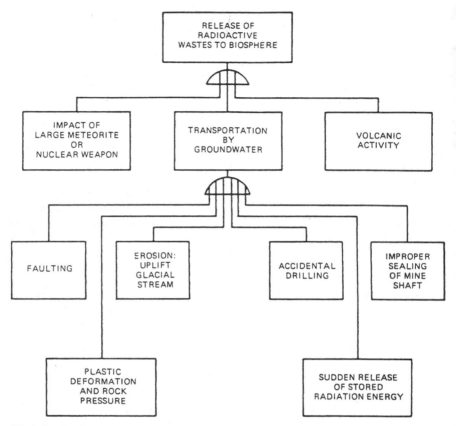

Fig. 3 Fault tree of salt mine used for storage of radioactive wastes (after closure of the mine). From McGrath (1974).

realize there were that many things that could go wrong."

The availability heuristic magnifies fears of nuclear power by blurring the distinction between what is remotely possible and what is probable. As one nuclear proponent lamented, "When laymen discuss what *might* happen, they sometimes don't even bother to include the 'might'" (B. Cohen, 1974). Another analyst has elaborated a similar theme in the misinterpretation of "worst-case" scenarios:

"It often has made little difference how bizarre or improbable the assumption in such an analysis was, since one had only to show that some undesirable effect could occur at a probability level greater than zero. Opponents of a proposed operation could destroy it simply by exercising their imaginations to dream up a set of conditions which, although they might admittedly be extremely improbable, could lead to some undesirable results. With such attitudes prevalent, planning a given nuclear operation becomes somewhat perilous since it requires predicting the extent to which the adversaries can employ their imagination" (J. Cohen, 1972).

An example of the politicized use of worst-case scenarios is a recent letter sent out by the Union of Concerned Scientists which contains the following statement:

"These are the facts: one accident from one plant could kill as many as 45,000 people, cause $17 billion in property damage, and contaminate an area the size of Pennsylvania."

Notice that no mention is made of the miniscule probabilities assigned to this worst-case scenario by those who developed it.

Whereas the availability heuristic implies that educational attempts may often backfire, another likely outcome is that information designed to educate people will simply have no effect. This consequence is implied by psychological research demonstrating that people's beliefs change slowly, and are extraordinary persistent in the face of contrary evidence (Ross, 1977). Once formed, initial impressions tend to structure the way that subsequent evidence is interpreted. New evidence appears reliable and informative if it is consistent with one's initial belief; contrary evidence is dismissed as unreliable, erroneous, or unrepresentative (Mazur, 1973). Thus, depending on whether one is predisposed in favour of nuclear power or against it, intense effort to reduce nuclear hazards may be interpreted to mean either that the technologists are responsive to the public's concerns or that the risks are great. Likewise, opponents of nuclear power may perceive minor mishaps as near catastrophes and dismiss the opinions of experts who rebut such claims as biased by vested interests. From a statistical standpoint, convincing people that the catastrophe they fear is extremely unlikely is difficult under the best condtions. Any mishap can be seen as proof of high risk, whereas demonstrating safety requires a massive amount of evidence (Green and Bourne, 1972). Nelkin's (1974) case history of a nuclear-plant siting controversy provides a good example of the inability of technical arguments to change opinions. Each side in that debate capitalized on technical ambiguities in ways that reinforced its own position. Similarly, the Swedish government's massive campaign to educate people about nuclear power and other energy sources showed that 10 or more hours of instruction had little influence on the attitudes of the 80,000 participants. The most significant effect was an increase in confusion and uncertainty about nuclear power, caused by an inability to resolve the conflicting opinions of technical experts (Nelkin, 1977).

Pathways Towards Acceptance

With all this working against it, what pathways might lead to widespread acceptance of nuclear power? Public response to X-rays and nerve-gas provides some clues. Widespread acceptance of X-rays shows that a radiation technology can be tolerated once its use becomes familiar, its benefits clear, and its practitioners trusted. Of course, X-rays are not perceived as potentially catastrophic, nor do they create the dread presently associated with nuclear power (see Figure 2).

Nerve gas also provides an enlightening case study. Few human creations could be more dread or more potentially catastrophic than this deadly substance. When, in December of 1969, the army decided to transfer nerve gas from Okinawa to the Umatilla Army Depot in Hermiston, Oregon, citizens of Oregon were outraged – except those in Hermiston. Whereas public opinion around the state was more than 90 per cent opposed, residents of Hermiston were 95 per cent *in favour* of the transfer, despite the warning that the fuses on the gas bombs deteriorate with age, but that the gas does not (*Eugene Register Guard*, Dec. 18, 1969 and Jan. 11, 1970). Several factors seem to have been crucial to Hermiston's acceptance of nerve gas. For one, munitions and toxic chemicals had been stored safely there since 1941 (so the record was good and presence of the hazard was familiar). Second, there were clear economic benefits to the community from continued storage at the depot of hazardous substances, in addition to the satisfaction of doing something patriotic for the country. Finally, the responsible agency, the U.S. Army, was respected and trusted.

These examples illustrate the slow path through which nuclear power might gain acceptance. It requires an incontrovertible long-term safety record, a responsible agency that is respected and trusted, and a clear appreciation of benefit. However, since people are generally willing to accept increased risks in exchange for increased benefits (Starr, 1969; Fischhoff *et al.*, 1978), a quicker path to acceptance might be forged by a severe energy shortage. Brownouts, blackouts, or rationing of electricity would undoubtedly enhance the perceived benefits from nuclear power and increase society's tolerance of its risks. A recent example of this process is the oil crisis of 1973-4 which broke the resistance to offshore drilling, the Alaska pipeline and shale oil development, all of which had previously been delayed because of their environmental risks. Such crisis-induced acceptance of nuclear power may, however, produce anxiety and stress in a population forced to tolerate what it perceives as great risk because of its addiction to the benefits of electricity.

What About Acceptance of Alternative Energy Systems?

Problems of public acceptance should be much less severe for non-nuclear sources of energy.   Fossil fuels and hydroelectric systems are familiar, common and appear to be perceived as less risky than they actually are.   Fossil, hydroelectric and solar energy systems have their roots in man's ancient past and work via mechanisms (combustion, water force, and sunshine) that are familiar, natural, and understood by all.

Many of the accidents and fatalities resulting from these systems involve individuals isolated spatially and socially from the rest of society.   Observation of the recent failure of the Teton Dam in Idaho suggests that even such catastrophic hazards as the collapse of a hydroelectric dam will quickly be forgotten – quite a contrast to the likely consequences of a reactor accident.

Conclusions

Development and regulation of nuclear power need to be based upon an understanding of the ways in which people think about risk and uncertainty. Our aim in this paper was not to document public opposition and fear of nuclear power, which are already well known, but to point out that this reaction stems in part from the recognition that there are important unresolved technical issues in the risk assessment process and in part from fundamental mental processes such as the use of imaginability and memorability as the basis for estimates of probability and frequency. Normal modes of thought, coupled with the special qualities of nuclear hazards that make them particularly memorable and imaginable yet hardly amenable to empirical verification, blur the distinction between the possible and the probable and produce an immense gap between the views of technical experts and a significant portion of the public.   This gap must be acknowledged and the difficulty of reducing it by appeal to reason or empirical demonstrations of safety must be recognized by planners and policy makers. Facing this problem means addressing some hard questions.   Does nuclear technology force us to make decisions that cannot be made well (or successfully) in a democratic society?   What kind of political institutions are needed to preserve democratic freedoms and insure public participation when decisions involve extreme technical complexity, catastrophic risk, and great uncertainty?[1]

---

[1] This article was written before the disastrous accident at the Three Mile Island reactor.   While it is too early to foretell the major consequences of this event, it will undoubtedly aggravate the disagreements about the risks from nuclear power.   Some experts will argue that even very improbable events do sometimes happen, and that their occurrence does not necessarily invalidate the previous assessment of their improbability. Others will point to the safety systems that did work as evidence of the effectiveness provided by overlapping systems. Still others will admit error but

Acknowledgements

This research was supported by the National Science Foundation under Grant ENV77-15332 to Perceptronics, Inc. Any opinions, findings and conclusions or recommendations expressed in this publication are those of the authors and do not necessarily reflect the views of the National Science Foundation. We are indebted to Drs H. Paschen and Chris Hohenemser for their constructive comments on an earlier draft of this manuscript.

References

Bronfman, L.M. and Mattingly, T.J., Jr (1976). *Nuc. Safety* 17, 539-549.
Bryan, W.B. (1974). Testimony before the Subcommittee on State Energy Policy, Committee on Planning, Land Use, and Energy, California State Assembly, Sacramento, California.
Cohen, B.L. (1974). *Bull. of Atom. Sci.* 30, 35-39.
Cohen, J.J. (1972). *In* "Risk vs. Benefit: Solution or Dream" (H. Otway, ed.), Report LA-4860-MS. Los Alamos Scientific Lab., Los Alamos, New Mexico.
Fischhoff, B. (1977). *Policy Sci.* 8, 177-202.
Fischhoff, B., Slovic, P., Lichtenstein, S., Read, S. and Combs, B. (1978). *Policy Sci.* 9, 127-152.
Green, A.E. and Bourne, A.J. (1972). "Reliability Technology". Wiley Interscience, New York.
Hohenemser, C., Kasperson, R. and Kates, R. (1977). *Science* 196, 25-34.
Holdren, J.P. (1976). *Bull. of Atom. Sci.* 32, 20-22.
Kasper, R.G. (1978). Presented at "Impacts and Risks of Energy Strategies: Their Analysis and Role in Management", Beijer Institute, Stockholm. (This Volume Part 1).
Lichtenstein, S., Slovic, P., Fischhoff, B., Layman, M. and Combs, B. (1978). *J Exp. Psych: Hum. Learning and Mem.* 4, 551-578.
Lifton, R.J. (1976). *Bull. of Atom. Sci.* 32, 16-20.
Lowrance, W.W. (1976). "Of Acceptable Risk". Wm. Kaufmann, Los Altos, California.
Mazur, A. (1973). *Minerva* 11, 243-262.
McGrath, P.E. (1974). "Radioactive Waste Management: Potentials and Hazards From a Risk Point of View". (Report EURFNR-1204 (KFK 1992)) U.S.-EURATOM Fast Reactor Exchange Programme, Karlsruhe, Germany.

---

say that the system can be improved so that such accidents will never again happen. Other experts and most of the public will interpret the Three Mile Island accident as evidence that the previous risk estimates were much too low. We expect the perception gap to increase. The slow path towards acceptance of nuclear power seems even less navigable than before. The need to address the hard questions appears even more urgent.

Melber, B.D., Nealey, S.M., Hammerslea, J. and Rankin, W.L. (1977). "Nuclear Power and the Public: Analysis of Collected Survey Research". (Report PNL-2430) Battelle Memorial Institute, Seattle, Washington.

Nelkin, D. (1974). *Bull. of Atom Sci.* **30,** 29-36.

Nelkin, D. (1977). "Technological Decisions and Democracy: European Experiments in Public Participation". Sage Publications, Beverley Hills, California.

Otway, H.J., Maurer, D. and Thomas, K. (1978). *Futures* **10,** 109-118.

Pahner, P.D. (1975). *In* "Risk-benefit Methodology and Application: Some Papers Presented at the Engineering Foundation Workshop" (D. Okrent, ed), 557-580 (ENG-7598). UCLA, Los Angeles, California.

Pokorny, G. (1977). "Energy Development: Attitudes and Beliefs at the Regional/National Levels". Cambridge Reports, Cambridge, Mass.

Primack, J. (1975). *Bull. of Atom. Sci.* **31,** 15-17.

Ross, L. (1977). *In* "Advances in Experimental Social Psychology" (L. Berkowitz, ed), 173-220. Academic Press, New York.

Slovic, P., Fischhoff, B. and Lichtenstein, S. (1979). *Environ.* **21** (3), 14-20, 36-39.

Starr, C. (1969). *Science* **165,** 1232-1238.

Tversky, A. and Kahneman, D. (1974). *Science* **185,** 1124-1131.

U.S. Nuclear Regulatory Commission (1975). "Reactor Safety Study: An Assessment of Accident Risks in U.S. Commercial Nuclear Power Plants" (NUREG-75/014). The Commission, Washington, D.C.

U.S. Nuclear Regulatory Commission (1978). "Risk Assessment Review Group Report to the U.S. Nuclear Regulatory Commission". (NUREG/CR-0400). The Commission, Washington, D.C.

Weinberg, A.M. (1976). *Amer. Sci.* **64,** 16-21.

Wilkes, J.M., Lovington, M., Horne, R., Pulaski, F. and Poole, R. (1978). Presented at "3rd Annual Meeting of the Society for the Social Study of Science". Bloomington, Indiana.

Zebroski, E.L. (1975) *In* Risk-Benefit Methodology and Application: Some Papers Presented at the Engineering Foundation Workshop" (D. Okrent, ed), 633-644. (Report ENG-7598). UCLA, Los Angeles, California.

PART 4:
RISK ASSESSMENT METHODS AND IMPLEMENTATION

INTRODUCTION

The theoretical structures and models for risk assessment are at odds with the application. Even specific methods such as formal cost-benefit analysis have been shown to be of doubtful utility by the authors in Part 4. Do we retreat and ignore risk-assessment, or do we attempt to find pragmatic approaches that can be of some use to decision-makers? Very few benefits are risk-free; inequitable risks are imposed on all of us to some degree; and some levels of risk, generally very low, are accepted by all societies. If pragmatic approaches are accepted, then processes are required that lead to risk levels being apportioned equitably throughout society when groups within it attempt to acquire benefits. Such a formulation is neither precise nor satisfying, but it opens the possibility of setting "acceptable levels of risk".

The purpose of the single overview paper by William Rowe in this final Part is to put forward a hypothesis summarizing public attitudes and reactions to issues and events containing identifiable risk elements. Thus, the approach is one of provoking discussion on what can incrementally help in the risk assessment process.

It will be immediately evident that no single method or approach is universally applicable for setting acceptable levels of risk, but that some approaches are effective. We should examine these in order to identify those elements which have led to success. Broad, fundamental questions are easily framed, but not easily answered. Finding out what has worked, and why, will at least enable us to ask the right questions in future.

What emerges from the paper is that the process of risk assessment may be more important than the methods used. The framework for such a process, is laid out, not as a specification for the process, but as a check-list of the items that must be considered and how they interact. This process may lead to the formulation of an "acceptable risk". If not, at least the right questions will have for reducing a whole range of disparate variables to common units and thus for conformably weighing the pros and cons against each other. In the first paper of this Part, David Pearce acknowledges the enthusiasm of the theorists; the technique is well suited to those contexts where costs and benefits are readily defined and where impacts do not extend into future generations. But practitioners have become aware of the doubts surrounding the use of cost-benefit techniques for issues which seem to transcend its framework. Many economists, including Pearce, now question how far the technique can go in the evaluation of energy futures beyond its obviously great value as a means of ordering and listing the pros and cons of investment options. As an example, he demonstrates that the use of discount rates in conventional cost-benefit analysis may have little relevance for energy futures. One can make such problems as those of the costs of nuclear-waste to future-generations, appear totally insignificant by the discounting principle, which holds that for the present society, a benefit or cost equal to one unit of money-value is worth much less in the future than it is now. This is only one of the many built-in assumptions which clearly indicate that cost-benefit analysis starts out from a number of moral assumptions or value-judgements. It is not so much that these moral assumptions cannot be changed to reflect more closely societal needs; they can – so long as these can be unambiguously identified. Invariably however, it can be argued, that because these value-judgements are implicit in the analysis, its use will only tend to obscure the ethical options which should be considered openly, particularly when debating energy futures.

Pearce's paper cautions the user that although cost-benefit analysis may be perfectible, it will only be practised perfectly in an ideal world peopled by perfect analysts. The second paper by Baruch Fischhoff examines this dilemma directly and adds considerable weight to the idea that however perfect the principle of cost-benefit analysis may become, it will only be as good as the fallible human-beings who practise it.

Fischhoff examines in some detail three important factors which limit the usefulness of cost-benefit analysis and related variants of this technique:

1) The unavailability of certain inputs necessary to the analysis

2) The inability of the analysts to assess the validity of their work, except at the conclusion of the recommended actions – when it is usually too late to change anything

3) The failure of the methods to address critical management issues, including the acceptability of the underlying political philosophy.

One or another of these points have been noted earlier, particularly by Pearce, Page, Slovic *et al.*, and below by Vedung and Staynes.

Fischhoff lists a basic set of criteria which such analyses should meet in order to gain widespread and wholehearted adoption and goes on to examine the effects of these limiting factors in relation to the basic criteria. Some of the problems that emerge are:-

1) the judgemental fickleness of "experts";

2) the assumption that the decisions are made for a single unified group in society which is not compatible with a pluralistic society;

3) the immutability of the analysis;

4) the tendency to believe that those values which cannot be measured by money, do not exist e.g. aesthetic and cultural values;

5) the danger of taking decision-making out of the hands of the public and giving it over to "expert-executives".

Against these difficulties, Fischhoff cites some assets of the technique, e.g. the explicit nature of its structure and inputs, which has probably raised the level of debate on those issues to which it has been applied and has broadened the understanding of even its harshest critics. And lest the reader be too disillusioned by this critical evaluation accorded to cost-benefit technique, Fischhoff enjoins us to consider very carefully what alternatives we have available to us before rejecting it completely.

As pointed out by Vedung in the last paper in this Part, some assessments of energy-risk stop after the first (technical) stage referred to above has been completed, leaving the second (policy analysis) stage completely to the decision-maker at the political level. Others try to cover the two stages sequentially to provide a series of policy options and their likely outcomes for decision-makers to work with.

The third paper in this Part, by Robin Dennis, is clearly concerned with the latter approach and overtly attempts to use the modelling process to elucidate policy-analysis and long-term management strategy. As pointed out in Part 1 by Harte, ecological considerations reach far into the future and Dennis claims that modelling can clarify such long-term environmental impacts. We can no longer plan the present without any concern for the future and hope to handle inconvenient environmental side effects by quick, technical fixes as and when they turn up. Dennis is emphatic that modelling should not *be* the analysis; but a good, explicit model can help to focus and clarify the analytical process, the act of modelling usually being much more important in this respect than the resultant model itself. This is a fundamental caveat stressed by Dennis as a result of extensive energy/environemtn modelling studies in Europe and USA.

Dennis also points out that there will always be some degree of conflict between socio-economic, energy and environmental issues and the trade-offs between these are normally left to completely informal processes, (discussion-meetings etc.), between the various interest groups. Yet exposing and clarifying the unwritten assumptions of each group in a systematic way provides more information of value to decision-taking than subjective discussion. Dennis examines the possible use of multi-attribute utility modelling to remedy this difficulty. Again, the actual process of using such models turned out to be the most important benefit to decision-taking and not the model itself.

The next two papers are concerned with the implementation of risk-assessment findings.

The fourth paper by Shelagh Staynes, illustrates the difficulties that could arise when governmental policy-makers and executive agencies are asked to respond to risk-analysis. As a case-study for this, the great complexity of the institutional arrangements that have grown up in the United Kingdom for the management of lead as a pollutant, both from motor-vehicle exhausts and from non-energy sources, is demonstrated by Staynes. However perfect the risk information may be, it does not necessarily follow that it can automatically be used for effective executive control. In a "closed-system" as found in Britain and some other European countries, with many "actors" and a complex web of interactions and constraints between them the point is clearly made that the outputs from risk-assessment should be developed and formulated with the realities of the management and control mechanisms very much in mind. The output from risk studies must be made to fit comfortably with the management system. Otherwise, unless mechanisms exist in govern-ment-administration whereby the multiplicity of "actors", special interests and management agencies, can be clearly identified, co-ordinated and made accountable, risk-assessments will be of doubtful influence in policy decisions and subsequent risk-managements. The conclusions for lead are not unique. Closely similar conclusions were made by Fischer in Part 1 for the control of oil-pollution in the North Sea, basically a much simpler administrative problem than lead pollution.

If Staynes' paper injects a note of sober realism into the discussion, the paper by Evert Vedung does nothing to dispel this mood. He somewhat realistically points out that perhaps the best way to gain an insight into the actual practice of risk-management is to study ways in which decisions about levels of acceptable risk have actually been made in the past. This involves the study of decisions *ex post facto*. Here the strategic and tectical considerations emphasised by Fischer and Staynes again come to the fore and are generalized into a series of concepts and hypotheses that attempt to explain decision-making in terms of the political pragmaticisms as outlined at the beginning of Part 1 by Carl Tham. Although many of these factors have been identified by other investigators, several postulates are new and should be validated against actual experience.

SOCIAL COST-BENEFIT ANALYSIS
AND NUCLEAR FUTURES

David W. Pearce

*Department of Political Economy,
University of Aberdeen, Scotland, U.K.*

If the advocates of nuclear power are correct, few countries, if any, seeking to industrialize or to sustain an industrial future, can afford to reject an energy future largely dominated by nuclear electricity. Quite simply, alternative energy sources – solar power, geothermal energy, wind and wave power – and massive energy conservation schemes, can not do much to meet the energy demands that a growth society requires, at least within the next half century. On this view, the benefits of nuclear power are at least approximated by the extra national income which can be brought about by its use.

One should perhaps stress that such a measure is the crudest possible approximation. Nearly everyone is now aware that Gross National Product (GNP) as conventionally measured has some peculiar features. As a simple example: if health risks, and therefore expenditures on prevention and care, are functionally related to income growth (the so-called diseases of the affluent society), then those expenditures will actually raise GNP even though welfare, in health terms at least, has been reduced. Nonetheless, for our purposes in making the "best" case possible for cost-benefit analysis, the approximation is perhaps acceptable. For attempts to modify the GNP measure to allow for such oddities, see Nordhaus and Tobin, (1972).

The costs of a nuclear future are less easy to delineate and less easy still to measure. These costs must include radiation hazards, the problems of disposing of high level radioactive waste to the ocean floor or in deep geological deposits, the problem of civil liberties in a world where potentially dangerous substances such as plutonium must be accounted for down to perhaps the last 100 kilograms, and the risks of nuclear proliferation. It is these latter aspects that make social cost-benefit analysis largely, though not entirely, irrelevant to the evaluation of energy

futures. If there is to be an informed and sensible evaluation
of nuclear programmes, quasi-quantitative techniques such as
cost-benefit analysis have little or no role to play. Once this
is accepted, it has many implications for *how* we decide on nuclear
futures both at the national and international level.

Social Cost-Benefit Analysis

The essence of cost-benefit analysis is simple. The object is
to establish all the costs and benefits relating to a given project
or programme and to select projects only if benefits exceed costs.
If more than one project has an excess of benefits over costs
and only some projects can be selected, projects may be ranked
according to their benefit-cost ratio. The guiding principle
must always be that a cost is anything that sacrifices some want
or incurs some loss of welfare and which would not have occurred
had the project in question not occurred. A benefit is any gain
in welfare brought about by the project and which would not
have occurred in its absence. Note that the concept of cost
incorporates the idea of a sacrificed benefit. The use of resources
to engage in research and development (R & D) for a fast-breeder
reactor therefore entails a cost which can be translated into
some foregone benefit, perhaps the results of R & D in alternative
energy resources. This foregone benefit, or "opportunity cost"
in the economist's language, may be approximately measured
by the money value of resources taken up in the fast breeder
R & D programme. Notice, too, that such definitions imply a
search procedure whereby one has to be sure that the analyst
has "captured" all the gains and losses attributable to the project,
or as many as he can. These effects may not cease at local or
national boundaries, and in the case of nuclear power they evidently
do not. Typically, however, cost-benefit analysis has not embraced
international repercussions: the nation has defined the geographical
unit. Having made this observation, there is nothing by way
of *conceptual* difficulty in extending the analysis to a global
or semi-global level.

Just as any analysis of nuclear power would have to be "stretched"
spatially, so it must be stretched in time. A road or an airport
has a limited life. It is either superceded by a new installation
and, of course, at the end of its physical life it can be taken
apart if it is not wanted. Even if it is left in a state of disuse,
its cost to a future generation, while finite, is unlikely to be
significant. With nuclear power we can make no such assumption.
The nature of the nuclear fuel cycle is such that spent fuel from
whatever kind of reactor, must be stored or reprocessed. Both
technologies have been practised and both will increase by orders
of magnitude as nuclear futures materialize. The debate as to
whether such fuels can be stored for ten or twenty years is of

no immediate concern for the current argument, although it is clearly vital if we wish to argue for a nuclear "pause" while we make a full assessment of the nuclear fuel cycle and its implications, such as that now taking place on an international basis (e.g. International Nuclear Fuel Cycle Evaluation programme – INFCE).   Whether spent fuel can be stored indefinitely (the "throw away" fuel cycle) is the subject of debate, and particularly so at the UK enquiry into the oxide fuel reprocessing plant proposal for Windscale.[1] But whether the spent fuel is stored for some period, or whether it is reprocessed to extract uranium and plutonium for further energy generating use, there are risks relating to low level radiation releases and to the storage and disposal of the remaining highly active waste.[2]   In the latter case, if vitrification (glassifying the waste) is successful, then generations to come will inherit that waste, deposited perhaps in deep mines or granite repositories. Since the half-life of these high level radioactive wastes can run to thousands of years, or, if certain technologies now under investigation are successful, hundreds of years, it follows that, unless there is an infinite assurance of zero risk, cost-benefit analysis would have to accommodate that risk in its "balance sheet".

As conventionally practised, however, such risks would attract zero cost in a cost-benefit study.   This occurs because of the use of a discount rate.   The rationale of discount rates is, on the face of it, simple.   A cost or a benefit in conventional cost-benefit analysis is, or should be, defined according to what individuals want.   The expression of a want implies a benefit and the money value of that want, measured by the individual's willingness to pay for the object of the want, is the sum that is entered in the cost-benefit balance sheet.   The expression of a "dis-want", a statement about an aversion to some effect of the project, would be measured by the sum required to tolerate its existence (the compensation measurement) or, less acceptably, by the sum the individual is willing to pay to avoid bearing the cost.   In the same way as preferences define costs and benefits, so individuals express preferences for the timing of those costs and benefits.   Strictly, a £1 benefit in the future is "worth" less to them than a £1 benefit now.   Since individuals "discount" the future the cost-benefit balance sheet incorporates a discount factor which scales down

---

[1] The full proceedings are given in the transcripts and submissions to the Local Public Inquiry into the proposal by British Nuclear Fuels Ltd to build a thermal oxide fuel reprocessing plant at Windscale in Cumbria.   Most, though by no means all, of the arguments are summarised in the "Windscale Inquiry Report" (1978).

[2] Reprocessing removes from the spent fuel plutonium, uranium and waste fissile material roughly in the proportions 97:1:2.

future gains and costs, so that, when summed across time, the
final result is a figure of present value or present worth.

The use of a positive discount rate has immediate implications
for our nuclear waste disposal example. Quite simply, for the
future generation a very small risk may have a large monetary
cost. Multiplying a large loss by a small probability can self-
evidently still leave us with a significant cost. But with a discount
rate that figure can be made to appear to totally insignificant.
For not only is one multiplying the (hypothetical) monetary loss
by a risk figure, one is multiplying that total again by a discount
factor. At an annual discount rate of 10 per cent the discount
factor in year 50 would be 0.009. That is, a cost of £100 million
to a future generation, with an attached probability of, say 0.1
would appear in the conventional cost-benefit balance sheet
as a cost of £0.09 million, i.e. only £90,000 compared to the expected
value to that generation of £10,000,000. It is easy to see that
cost-benefit, as conventionally practised, can make the nuclear
waste disposal problem vanish by analysis.

Before investigating these issues in a little more detail, let
us consider one more feature of cost-benefit analysis. In a world
of finite resources, we cannot provide sufficient to satisfy the
wants of everyone. Every choice therefore implies an opportunity
cost. The logic of this is inescapable. To have one thing we
go without the other. Suppose now we can estimate the money
benefits of nuclear power. Suppose too that we can estimate
some of the costs but some are left as "intangible" items. If
we choose nuclear power it is easy to demonstrate that we shall
have implied a value for the intangible risks. The same is true
if we decide against nuclear power (or some aspect of it).

Quite simply, if the benefits are X and the known costs are Y,
then the decision to proceed *implies* that the unknown costs must be
less than X-Y. In the same way, if we reject the proposal, we imply
that the unknown costs are greater than X-Y. We have not produced
an exact money value, but we have secured an upper or lower bound.
Great play is made of this logic by economists in pointing out that,
whether we know it or not, we always attach money values to
outcomes. It is often used as a justification for cost-benefit analysis.
In reality, while the logic has its uses, it cannot be used for such a
purpose. In the first place, the implied values are those of the person
making the decision and do not relate back to the value judgement we
noted earlier, that it is the values of the affected individuals that are
assumed to matter. What is happening is that we are falsely
aggregrating values formed in one context to values obtained in
another. Second, and more important, the implied values come about
because of the decision and have not been obtained as exogenous data
prior to the decision. Yet the point of cost-benefit is to use the
exogenous data to guide or decide upon the project worth. The use of
implied values reverses this logic since the values emerge as a result

of the decision and do not form an input into making that decision. We should not therefore be over-impressed by the economist's sleight of hand in these matters.

Values and Cost-Benefit

Many economists have become aware of the doubts surrounding the use of cost-benefit techniques for issues which seem to transcend its framework. In truth, the cost-benefit paradigm is suited only to those contexts where the benefits and costs are readily defined and where there is at least a *prima facie* case for expecting that some approximation of individuals' value can be obtained. There must be doubt as to whether it is suited to any context in which there are effects which are long-lived and potentially significant. The example of persistent ecosystem damage may be cited (Pearce, 1976), as might the examples of lead or cadmium (Pearce and Nobbs, 1976), or, if scientifically proven to be a problem, the accumulation of carbon dioxide in the atmosphere. Note that, if this is correct, we have no more reason to espouse cost-benefit techniques for the evaluation of a coal-investment programme, since the use of coal in power generation contributes significantly to CO_2 accumulation and adds to heavy metal emissions. It is no part of this paper to argue that the environmental aspects of nuclear power are so special that it should be singled out on this basis. It will be suggested later that there are other aspects which *do* make it unique in terms of its costs and which therefore warrant special attention.

It has been suggested by one eminent economist (Kneese, 1977), that cost-benefit is inapplicable to nuclear issues because a commitment to nuclear power involves unavoidable moral issues. This is partly correct, but overlooks the fact that cost-benefit is already morally loaded. If this is correct, it seems odd to suggest that cost-benefit cannot handle moral problems when it already incorporates an ethic of its own. It is already, as it were, a moral science. To demonstrate its moral premises is not difficult given our earlier discussion of how, in outline, cost-benefit proceeds. Essentially, three value judgements are involved. First, it is preferences that count. The aim is therefore to seek a monetary valuation for those preferences. Second, because of the use of valuation procedures based on willingness to pay or compensation required, the valuations are income-biased. Essentially, they reflect the income distribution that *exists*. Yet this is tantamount to saying that that distribution is "optimal". The second value judgement can therefore be stated as: an individual's valuations should be weighted by the purchasing power of that individual. The third judgement may be subsumed under the first but it is useful here to separate it out. Whose preferences are to count? We have already drawn attention to the oddity of sometimes using the values of individuals and other times using the values of decision-makers. More important perhaps is the fact that, as we saw, the discount rate downgrades the values of future generations and weights heavily the values of current generations. In short, we may reasonably state the third judgement as: the values of the current generation should generally prevail (see Nash et al., 1975, for a fuller discussion).

Now, without elaborating on this issue too far, since it is a complex and far-ranging problem we may note that there is nothing "sacred" about any of these judgements. We may wish to overrule individual preferences in favour of paternal values. We may not regard the existing distribution of income as optimal and may prefer to adopt some criterion of justice or fairness, such as that advocated by Rawls (1972). And we may certainly not wish to ignore future generations' costs and benefits as seen from *their* standpoint in time. In respect of the last issue it may seem sensible to ignore future generations' views since we cannot know what they are. In the context of nuclear power, however, this is a peculiar argument since many of the decisions being made now are designed to supply an energy future at least 25 years ahead and, generally, 50 years ahead. That is, the decisions being made *now* must be *for* future generations.

An interesting dilemma emerges in such a context. If future generations do opt for a growth economy ethic it may well be the case that they will not be able to secure it without fast breeder reactors. If so, failure to build the reactors, or their prototypes now, will deny future generations the benefits that this kind of energy can bring. If future generations do not opt for such an economy, they can of course close down whatever reactors there are and settle on a lower economic growth path. Yet in doing this they will still have inherited the waste disposal problems generated by the decisions made now, together with an inventory of plutonium they may not want.[3] This dilemma is mentioned simply because it does emphasize the ethical issue rather well: to opt for nuclear futures or not to opt for them may both be decisions which impose irreversible consequences on at least one generation, and probably many more. In the United Kingdom, for example, consideration is currently being given to construction of the first commercial size (1200 megawatts) fast reactor. If, as seems likely, permission is given to commence construction in the 1980s, it cannot come "on line" until about 1990 and, since its alleged function is to demonstrate the viability of fast reactors to the purchasers (the electricity utilities), it is doubtful if a second or third fast reactor would be on stream much before the year

---

[3] Note that this inventory has little to do with fast reactors which can be set so as to "incinerate" plutonium – i.e. act as waste disposal plants. The major part of the inherited inventory comes from reprocessed spent fuel from thermal reactors.

2000.   Since no-one apparently argues that the fast reactor will
definitely be needed, its justification rests on a probabilistic
argument to the effect that the generations of twenty or more years
in the future must have a proven technology to hand if they want it.
If this is correct, current energy planners are revealing a marked
degree of altruism, although it would be naive to believe that current
rewards are not also relevant to the decision. It is considerations
such as these that have led to some interesting proposals from
economists as to how we might modify our approach to such
commitments. A brief note about the work of Page (1977) is perhaps
in order. Essentially, Page argues that while nuclear futures may
offer the most immediate prospect of benefits, alternative energy
futures may operate at lesser unit social cost at much later dates. If
the option to adopt nuclear power for a limited period followed by
some alternative energy source is not available, perhaps because a
nuclear programme precludes the necessary R & D to secure the
benefits of the alternative programme, then we have somehow to
weigh up the short-term benefits to, say, the next two generations,
against the longer term benefits of the alternative programme to
more distant generations. Discounting procedures would automatically favour the nuclear programme, as we have seen. Page suggests
instead that we should take a longer view and adopt an "almost
everywhere dominant" standpoint.   That is, we should consider
generations as if they had equal rights in a kind of intertemporal
assembly now (a kind of Rawls' principle applied through time) and
seek the outcome which benefits the maximum number of
generations. This summary hardly does credit to Page's work, but it
will be seen that his principle would favour sacrificing the benefits of
nuclear power now and for the next generation or so in order to
secure the intergenerational benefits which come about because of
the absence of a discount rate. Of course, there are many pitfalls,
not least the argument that sacrificing nuclear power now may mean
sacrificing economic growth and thus reducing the chances of
securing investment in longer term alternative energy futures, but
these are the subject of a separate paper.

A Nuclear Fuel Cycle Example

The preceding sections have attempted to show what cost benefit
does and why it is unavoidably value-loaded. What is important
is that our argument suggests cost-benefit analysis is suited to
"moral" issues. Simply because there are ethical problems is
not a reason for rejecting cost-benefit analysis and it is here
that we take issue with Kneese (1977). Essentially, if cost-benefit
is already value-loaded its outcomes can be altered to reflect
different value assumptions.

All this said, what this argument preserves, if it is accepted,
is the *conceptual validity* of cost-benefit analysis. Now the practical

nature of cost-benefit must be considered.   While not going so far as to say that the most fervent advocates of cost-benefit are those who have never tried to practise it, it does seem that economists and other advocates of cost-benefit techniques severely understate the applicability of a conceptually self-contained technique.   The reasons for this are interesting in themselves and cannot be dwelt upon here, save to say that denial of the enormous gap between theory and practice most probably stems from the very philosophy of economics as a social science with its idea that any social process can be "modelled" and then estimated.

The suggestion that there are practical difficulties ranges far wider than any empirical study one might attempt on a nuclear fuel cycle.   It extends to whole areas of environmental study where practitioners claim to have secured "scientific" results for cost-benefit studies of air and water pollution and noise nuisance. The economics is complex but it is certainly arguable that the money valuations for cleaner air and less noise, secured in dozens of studies, are without meaning.   Harris (1978) and Mäler (1977) have some interesting observations to this effect.   It is very arguable that cost-benefit analysis has had little significant success in environmental analysis.

However, let us concentrate on a case study in the nuclear context to see what cost-benefit might have to say.   Consider the decision as to whether or not to build an oxide fuel reprocessing plant at Windscale in Cumbria, England.   This decision was publicly debated at a public inquiry in 1977 and the Report of that Inquiry was published in 1978.   Fuel from the UK's first generation nuclear stations, the Magnox stations, is already processed at Windscale and will continue to be.   The issue in question was the proposal to build a major plant to accommodate the reprocessing of spent oxide fuels from the programme of advanced gas cooled reactors (AGRs) and whatever reactor system is opted for in the 1980s (almost certainly a water reactor system).   As such, the plant implied an option for a "recycling" fuel cycle, in which uranium and plutonium are separated out from spent fuel for use in thermal or fast reactors.

The issues that arose were:

1) Whether the reprocessing plant (THORP which stands for "thermal oxide reprocessing plant") was needed (a) as a waste disposal plant and/or (b) as a means of inventorying plutonium for future fast reactors or as a "one-off" fuel for thermal reactors.

2) Whether THORP added to low-level radiation levels to an extent that could be considered unacceptable.

3) Whether THORP contributed significantly to the risk of accident with consequent serious radiation exposure to the work-force and/or local (or wider) populations.

4) Whether THORP entailed any infringement in terms of

civil liberties both to workers in the plant and persons outside.

5) Whether THORP contributed to the chances of proliferation of nuclear weapons by risking the loss of fuels which could be used for the manufacture of nuclear weapons from civil nuclear fuels.

6) Whether vitrified waste posed threats to future generations.

We may briefly overview each issue to see whether there is anything *in practice* that a cost-benefit study could do.

Issue 1: The Need

In any cost-benefit study one would expect to see the private costs and benefits of the project in question clearly expressed so that external costs and benefits could be added to form the overall social cost/benefit picture. The costs of THORP seem to be known within some range of error and economists are used to the idea of building in some cost escalation factor. But what are the benefits? Much depends on how the plant is perceived. If it is a waste disposal plant it seems proper to add the cost of running the plant (and associated plant such as one for vitrification of high level wastes) to the cost of nuclear electricity generation. One might then proceed as normal in cost-benefit approaches and look at the demand curve for electricity and how it might shift over time, and attempt to measure the welfare gains of having this electricity. In itself, it would not matter if THORP made a profit or not since it would be inseparable from electricity generating stations and its costs could be apportioned to any stations generating oxide fuel-waste for treatment.

The issue becomes more complex if THORP is seen as something more than a waste-disposal plant. Reprocessing involves taking the spent fuel and extracting from it uranium, plutonium and waste. Clearly, the uranium and plutonium can be re-used. The waste must be disposed of and it is this that causes the intergenerational disposal problem discussed earlier. Although there is, strictly, no such thing as a world price of uranium, it is reasonable to suggest that one could value the uranium so recovered. One might want to "shadow price" this fuel because, in the UK situation, the alternative is to import it at possibly great cost (depending critically on the politics of uranium in the future). There is no market at all in plutonium, although a quasi-market will emerge as THORP and similar plants reprocess foreign fuels (THORP will reprocess Japanese oxide fuel-waste, for example). But, again, one might consider the uranium equivalent of the plutonium and attach a price on this basis. Here again, however, much depends on where it is used since plutonium is most efficient inside fast reactors and is widely considered to be of little value in mixed fuels for thermal reactors.

Now, what this means is that THORP can be seen as a plant

which is designed to stockpile plutonium in the UK for use in
fast reactors should they be needed in, say, 2010 or even 2025
or 2050.   Such a policy makes more sense if there are constraints
on uranium supply to the UK in that period.   The benefit of THORP
may therefore be calculated in terms of the value of the plutonium
as an energy resource, approximated perhaps by a formula for
price such as that suggested above.   Cost-benefit principles would
seem to be preserved even if we view THORP in this fashion.
In reality, however, the situation is more complex.   For, if there
are any uranium constraints, the UK's thermal reactor programme
will be faced with difficulties, even though uranium may be a
small component of the total fuel cycle cost.   Those difficulties
may be such that some planned rate of GNP growth cannot be
sustained.   The benefit of THORP would be that it would have
stored plutonium for use in fast reactors which do not rely on
imported uranium for their capability.   In short, should the benefit
of THORP be something more than the price of the fuel recovered
by it?   Should it be some probability of securing a growth rate
in GNP that might not otherwise occur?   Without some knowledge
of the probabilities, covering a highly political market (uranium)
and some certainty about the GNP/energy relationship it is unclear
what order of magnitude the cost-benefit analyst should enter
into the picture.   Note too that it is not even clear if an analysis
of THORP *alone* would make any sense.   For, while it is true
that a number of fast breeders in the UK could be sustained through
the use of spent and reprocessed Magnox fuel (which does not
go through THORP), it is far from clear what THORP is for if
it is not to insure against a future in which uranium is constrained
in supply and fast reactors become "necessary".   There are clearly
formidable *practical* problems in separating out the two investments
and in gauging the value of such an insurance policy.

Issue 2: Low Level Radiation

Turning now to the side effects of THORP, any nuclear installation
adds to the level of radiation in the atmosphere or sea.   The
appropriate question then is whether those levels are "acceptable".
At the Windscale inquiry there was little dispute as to what additional
amounts of radiation THORP would give rise to.   The dispute
was largely over what the effects of low level radiation are.
The differences of opinion among international expets were substantial.
As such, there was a range of estimates of "probable" death to
work with, although the Inquiry Report rejected the evidence
to the effect that the worst levels were the correct ones and
opted instead for the lesser levels.   These implied exposure to
workers of about 1 rem per annum (one-fifth the current accepted
maximum) and to the most exposed populus of about 50 mrem
per annum (one-twentieth of a rem).   Apart from medical exposure

(X-rays) the current acceptable level for the general public is 500 mrem per annum so that THORP would *add*, at most, one-tenth of the maximum dose to the existing levels.

The difficulties for the cost-benefit analyst are again complex. Apart from having to seek a value for a human life, an exercise repeatedly investigated in the economics literature with a singular lack of conviction about any of it, these losses are probabilistic both in time and according to which expert one believes. Assuming some valuation for a single life could be obtained, however, the analyst could legitimately multiply that valuation by a range of probabilities and discount the result to secure a present value. It is arguable, then, that the fabric of cost-benefit could just be preserved in such an exercise. (Note, however, that the idea of discounting is almost inconsistent with the argument that THORP exists to supply fuel for future generations: strict application of a discount rate might lead one to think THORP would not be built at all).

Issue 3: Accidental Radiation Discharge

Very much the same analysis as above would apply for accidents. That accidents occur in nuclear installations is not now a challengeable fact. The suspicion that they are not all publicly recorded is unavoidable. Nonetheless, the problem that arises in applying cost-benefit techniques is that the probabilities of accidents are unknown for unknown technologies. No commercial sized reprocessing plant exists on the scale proposed for THORP nor does any commercial sized fast reactor exist. Hence the issue is not so much one of calculating the damage from a hypothetical release of radiation as of calculating the risk of accidents. In a 26 year period (1950-76) British Nuclear Fuels Ltd, who would operate THORP and who currently reprocess Magnox fuel at Windscale, reported 177 accidents (or, rather, reported them at the Inquiry in 1977), between 6 and 7 per year. However, these range from "innocuous" episodes to more serious episodes involving the contamination of 35 workers. It remains the case that extrapolation from this experience to the THORP plant is not one that can be based on objective probabilities. Again, the cost-benefit analyst has a difficulty in applying his principles to the practice. Arguably, the risks are so small that they could not alter any outcome in any event.

Issue 4: Civil Liberties

Once past issues 1-3, it becomes clear that cost-benefit analysis has little or no role to play in an assessment of investments such as THORP. It was accepted in the Inquiry and in the Report that civil liberties would be infringed. The types of security measures involved are fairly obvious: workers at the plant would have to be checked in case of attempted sabotage or theft of materials, including plutonium; those responsible for the transit of plutonium or waste would also have to be checked; there would continue in existence the private police force already established under Act of Parliament for the Atomic Energy Authority; there would be surveillance and investigation of parties thought to be opposed to nuclear power in such a way that their actions might be subversive. While the extra loss of liberty from *one* plant might be tolerable, it is legitimate to ask whether that loss would be acceptable for a nuclear programme, especially one where plutonium required regular transport from one place to another, including across national frontiers (Justice, 1978).

Now, it is no part of this paper to argue whether these infringements are acceptable or not especially when compared to the suggested benefits from nuclear power. The issue at stake here is whether they can be integrated into the cost-benefit calculus. The answer must be negative since it is quite unclear what answer we would expect to the question: "What are you willing to pay to avoid some (unknown) probability of increased security surveillance of innocent parties?" or the same question concerning surveillance of parties which are involved in the generation of nuclear power, and so on.

Issue 5: Nuclear Proliferation

Here the cost-benefit analyst faces the biggest challenge. Arguably, technologies to reduce terrorist use of plutonium for the manufacture of a nuclear weapon will emerge such that the chances of terrorist success are reduced to a minimum. Suggestions that fuel being transported be irradiated so as to make it unattractive fall into this category. Other issues relate to the possibility that such fuels could be diverted to countries without their own weapons capability but who might use diverted fuels, reprocess them and achieve some military nuclear capacity. What the risks are seems to be unknown. Much depends on the speed with which anti-terrorist technologies emerge. Much depends on just how realistic it is to expect any non-signatory of the Nuclear Proliferations Treaty to be *able* to reprocess fuels for military use, given that reprocessing is itself a complex and fairly large scale technology. But the very existence of a United States policy on non-proliferation through the non-export of nuclear fuels and the discouragement of reprocessing elsewhere lends support to the idea that the risk is positive. If so, that risk has to be multiplied by the risk that a nuclear capability would be used and the risk that its use would escalate to a war of wider proportions, then, whatever the value of human life, it is unclear in the extreme what the cost-benefit analysts have to say.

Issue 6: Future Generations

If there is to be a nuclear programme, spent fuels necessarily arise. In some countries, e.g. Canada, no reprocessing facilities are thought necessary, for the time being anyway, because of success in storing spent fuels. In the UK there are doubts as to whether the cladding of the spent fuel can withstand corrosion beyond, say, 10 years although some oxide spent-fuel must be stored for longer than that until THORP is built. Whatever the truth of the storage versus reprocessing argument (and reprocessing is, as we have seen, inextricably bound up with a resource-recovery argument), reprocessing does result in a high level radioactive waste which must be stored in special containers. Those containers may eventually be dumped on the ocean floor or in deep geological formations. The most promising technology is vitrification, the glassification of waste and its containment in concrete blocks. Experience with vitrification is limited and it would be incorrect to suggest that it is proven under the circumstances envisaged for the waste from THORP. Certainly, it seems clear that its behaviour must be monitored over very lengthy periods of time and, if so, it should be "recoverable" for further treatment if necessary. How far this constitutes a serious problem for future generations is debatable. Certainly the more alarmist worries would seem unfounded, but too many questions remain for it to be declared a "no-risk" policy. As we noted, however, that risk may accrue to generations whose voice is not heard and yet who inherit both the benefits and costs of nuclear power. It is far from clear that cost-benefit analysis can handle this problem.

Conclusions

Cost-benefit analysis has a distinc use in "ordering" thoughts, in running up a list of the pros and cons of any investment. Since many decisions are taken without even that much analysis taking place it is the case that cost-benefit analysis has a valuable function. How far it can go beyond this is open to question. It has undoubted use in contexts where the costs and benefits are well-defined and is more likely to succeed where they are localised rather than globally dispersed. Once issues such as investments in nuclear power or parts of a given nuclear fuel cycle are considered, however, its virtues are dissipated. It is far from clear that single investments can be divorced from whole programmes, as was suggested in the case of the Windscale/THORP example. It is also essential to face up to the issue of valuing human life and dealing with disputed probabilities. Perhaps more important, in a context where alternative technology is thought incapable of meeting energy demand and where energy use and GNP are highly correlated, the issue arises of what measure of welfare loss to use if a nuclear investment is *not* undertaken.   Arguably, one might estimate how much energy is foregone, translate this into foregone GNP through some projected energy coefficient, and use this as a first approximation.   Alternatively, one might look at the cost of trying to meet that energy demand by some other means. Both approaches are hazardous if only because of their location in time: one is projecting changes far into an unknown future.

The overwhelming reasons why cost-benefit ceases to function however, arises from the fact that, while many of these issues arise in other investments, e.g. an expanded coal programme, two issues are generally peculiar to nuclear power.   The first is the issue of the liberty of the individual.   The second is the risk of military/nuclear weapons proliferation and the associated risk of war on any scale.

Ultimately, the choice is an ethical and political one. We have suggested that cost-benefit itself is a moral evaluation technique since it presupposes value judgements which may or may not be acceptable.   But it would be a serious error to conclude from the common element of ethics that decisions about a nuclear future can be made with the assistance of cost-benefit analysis. Indeed, it can be argued that because cost-benefit embraces certain value judge-ments (although it need not embrace any particular set), its use will merely obscure the ethical *options* that should be considered when debating the desirability of a nuclear future.

References

Harris, A. (1978).   "Valuing Environmental Amenity: A Critique of the House Price Approach". (occasional paper 78-01). Department of Political Economy, University of Aberdeen.

Justice, (1978). "Plutonium and Liberty". Justice, London.

Kneese, A.V. (1977).   Benefit Cost Analysis and the atom.   *In* "Economics in Institutional Perspective" (R. Steppacher, ed.). D.C. Heath and Company, Lexington, Massachusetts.

Måler, K.G. (1977). A note on the use of property values in estimating marginal willingness to pay for environmental quality. *Journal of Environmental Economics and Management* **4,** 355-369.

Nash, C.A., Pearce, D.W. and Stanley, J.K. (1975).   An evaluation of cost-benefit analysis criteria.   *Scottish Journal of Political Economy* XXII, 121-134.

Nobbs, C. and Pearce D.W. (1976).   The economics of stock pollutants: the example of cadmium.   *International Journal of Environmental Studies* **8,** 245-255.

Nordhaus, W. and Tobin, W. (1972). Is growth obsolete? *In* "Fiftieth Anniversary Colloquium V," National Bureau of Economic Research. Columbia University Press, New York.

Page, T. (1977).   "Conservation and Economic Efficiency".   Johns

Hopkins Press for Resources for the Future. Baltimore Maryland.

Pearce, D.W. (1976).   The limits of cost-benefit analysis as a guide to environmental policy. *Kyklos* Fasc. 1, Vol. **29,** 97-112.

Rawls, J. (1972). "A Theory of Justice". Oxford University Press, Oxford.

Windscale Inquiry Report, Vol. 1. (1978).   Her Majesty's Stationery Office, London.

BEHAVIOURAL ASPECTS OF
COST-BENEFIT ANALYSIS

Baruch Fischhoff

Decision Research, A Branch of Perceptronics,
Eugene, Oregon, U.S.A.

Cost-benefit analysis asks whether the expected benefits from a proposed activity outweigh its expected costs. Prominent variants are cost-effective analysis (comparing the relative benefits of alternative actions), risk-benefit analysis (in which the major costs are risks to life, limb or property), and decision analysis (in which the need to reduce all consequences to monetary units is avoided and the role of uncertainty is made more explicit). For further details on these procedures, see Keeney and Raiffa (1976); Layard (1974); Stokey and Zeckhauser (1978); Raiffa (1968).

The expected cost of a project is determined by enumerating all aversive consequences that might arise from its implementation (e.g., increased occupational hazard), assessing the probability that each will occur, and estimating the cost or loss to society should each occur. Next, the expected loss from each possible consequence is calculated by multiplying the amount of the loss by the probability that it will be incurred. The expected loss of the entire project is computed by summing the expected losses associated with the various possible consequences. An analogous procedure produces an estimate of the expected benefits.

Although based on an appealing premise and supported by a sophisticated methodology, these procedures have a number of characteristic limits on their usefulness as management tools. These limits arise when the mathematical formalisms confront the fallible individuals who must conduct, accept, or implement them. Be they technical experts, lay interveners or government regulators, these individuals all have, to some extent, limited capacity to process technical information, restricted resources to devote to the project at hand, irrational apprehensions about its consequences, intransigent prejudices about the facts of the matter, ulterior motives, and incoherent and unstable values on critical issues. Deliberately or inadvertently, these human properties tend to foil the best-laid plans of cost-benefit analysts or the purveyors of almost any other scheme for managing technologies in society.

In considering the viability of any approach to management, it seems to be useful to have in mind a set of desiderata such as that in Table 1. Although this list is by no means exhaustive, it does at least

TABLE 1

Criteria for Evaluating Approaches to Managing Technologies

Logically sound

Implementable

Politically acceptable

Respects institutional constraints

Open to evaluation

Creates no side effects

Promotes effective long-term management

---

suggest some of the criteria a proposed approach must meet in order to merit wholehearted adoption. No proposal achieves perfect marks in each category. Analysing the strengths and weaknesses of each proposal can show where it needs improvement and for which problems it might be most suitable.

Rather than consider cost-benefit analysis from each of these perspectives, some of which are covered by other contributions in this book, I will discuss the effects of three sets of limits on members of this family of analytical procedures. One set of limits is imposed by the unavailability of necessary inputs to the analysis. Without those inputs, the implementability of cost-benefit analysis must be seriously questioned. A second set of limits comes from the inability of the analysts to assess the validity of their work and incorporate that assessment into their guides to action. The absence of such appraisals creates problems for the logical soundness of such analyses and their ability to protect the public from unanticipated side effects. The third set of limits comes from the failure of these methods to address critical management issues. These include the acceptability of the political philosophy underlying the procedures and the feasibility of implementing their recommendations.

Availability of Inputs

Performing a full-dress analysis assumes, among other things, that:

1) all possible events and all significant consequences can be

enumerated in advance;

2) meaningful probability, cost and benefit values can be produced and assigned to them; and

3) the often disparate costs and benefits can somehow be compared to one another.

Unfortunately, some of these tasks cannot be completed at all; while for others, the results are hardly to be trusted. Despite the enormous scientific progress of the last decade or two, we still do not know all of the possible physical, biological and social consequences of any large-scale energy project. Where we know what the consequences are, we often do not, or cannot, know their likelihoods. For example, although we know that a reactor core melt-down is unlikely, we will not know quite how unlikely until we accumulate much more on-line experience. Even then, we will be able to utilize that knowledge only if we can assume that the system and its surrounding conditions remain the same (e.g, there will be no changes in the incidence of terrorism or the availability of trained personnel). For many situations, even when a danger is known to be present, its extent cannot be known. Whenever low-level radiation or exposure to toxic substances is involved, consequences can be assessed only by somewhat tenuous extrapolation from the consequences of high-level exposure to humans or low-level exposure to animals (Harriss et al., 1978).

In all these cases, we must rely upon unaided human judgement to guide or supplant our formal methods. Research into the psychological processes involved in producing such judgements offers reasons for pessimism. A rather robust result is that people have a great deal of difficulty both in comprehending information under conditions of complexity and uncertainty and in making valid inferences from such information. The fallibility of such judgement stems in part from the counter-intuitive nature of many probabilistic processes, in part from the lack of hands-on experience with low-probability and high-consequence events, and in part from the mental overload created by many problems (Slovic et al., 1977).

The failings of our intuitions are shown in persistent tendencies to neglect various kinds of normatively important information, such as population base rates (indicating how common a particular event is), sample size (indicating how reliable evidence is) and predictive validity. Other kinds of information are attended to, but given inappropriate interpretations. People tend to be more confident making predictions on the basis of redundant information than with independent information (although the latter has greater predictive validity), they readily find interpretable patterns in random sequences, and they assume that more information guarantees better performance even when it only generates confusion. When asked to synthesize information from their experience, people tend to misjudge the risks to which they are exposed, they remember themselves to have been more foresightful – and others to have been less foresightful – in past judgements than was actually the case, and they sometimes persevere in erroneous beliefs despite mounting even overwhelming, contrary evidence, (Tversky and Kahneman, 1974; Fischhoff, 1975; Ross, 1977).

To the best of our understanding, these judgemental biases do not reflect folly, but the occasional (or frequent) inadequacies of people's best attempts to muddle through difficult problems. Often these attempts reflect the use of judgemental heuristics or rules of thumb that embody moderately valid appraisals of regularities in the world around us. In situations allowing a trial-and-error, successive correction approach to decision problems, these heuristics often work pretty well. However, where we (as planners or consumers) must get our decisions right the first time or suffer severe consequences, the limits of these heuristics may spell real problems.

These problematic tendencies are typically observed in situations where people's sole motivation would seem to be making the best, most "objective" judgement possible. Such situations are designed to avoid the additional problems engendered by the colouring of judgements by wishful thinking, self-serving motives, selective attention, cognitive dissonance and the like. Although the evidence is sketchy, there is at the moment no good empirical reason to believe that these problems are appreciably reduced when the judgement in question carries high (personal or societal) stakes or when the judge is a substantive expert forced to go beyond the available data and rely on intuition, (Slovic et al., 1977; Fischhoff, 1977).

Such results provide strong evidence for using formal methods for producing and combining information whenever possible – and for treating the results of such analyses with considerable caution because of their inevitable judgemental component.

Once the consequences have been enumerated and their likelihood assessed, a price tag (in dollars or utiles) must be placed on them. When it comes to trade-offs between deaths today and deaths in the future, between sterility and black lung disease or between profits and lives, both the exigencies of our political processes and the indeterminate nature of cost-benefit logic force us to ask people for their opinions.

Such questions of value would seem to be the last redoubt of intuitive judgement. Unfortunately, however, subtle changes in how questions are posed can have a major impact on the opinions elicited. Worse yet, in situations where alternative questioning procedures elicit different preferences, the normative theory often offers no guide as to which of the different jdugements is to be preferred. When people's judgements show this sort of lability, the method may become the message, leading to decisions not in the decision-makers' best interest, to action when caution is desirable (or the opposite), or to the obfuscation of poorly articulated views, (Fischhoff et al., in press).

Many of these effects have been known since the beginnings of experimental psychology in the mid-1800s. Early psychologists concerned with the relationship between sensations and judgements about them found that both the threshold for discerning a sensation and the threshold for discriminating between two sensations depended on a variety of subtle aspects of how stimuli were presented and how responses were elicited. Different judgements were attached to the same stimuli as a function of whether those stimuli were presented in ascending (increasing on a physical continuum) or descending order, whether the set of stimuli was homogeneous or diverse, whether particular regions on the continuum were densely or sparsely represented, whether sequentially presented stimuli were relatively similar or disparate, whether values near the threshold of detection were included or not, and whether the respondent made one or many judgements. Even when the same presentation was used, different judgements might be obtained with a numerical or comparative (ordinal) response mode, with implicit instructions motivating speed versus accuracy, with a bounded or unbounded response set, with small or large numbers (subsequently normalized), or with verbal or numerical labels. The instability of judgement is heightened by the fact that perception is inherently accompanied by some random error and by idiosyncratic tendencies such as fatigue, locking in on stereotypic ways of viewing a problem, second-guessing the elicitor (what am I supposed to say?) and linking variables that should be independent (halo effects).

All of these problems emerge when people are questioned about their values or preferences. The elicitor must decide how many questions to ask and how to word them; how many alternatives to consider and in what order to present them; what response format to use and how much time to allot for it. The preferences expressed will reflect in part the respondent's true beliefs, in part the method used to uncover them. Indeed, no decision is so clear cut in its options, events and attributes, no respondent is so mechanical, that these problems can be avoided entirely.

Particular kinds of lability seem to emerge when people are asked about value issues of the sort raised by proposed energy strategies. For such new and complex issues, with subtle interactions and gargantuan effects, people may have no articulated preferences. In some fundamental sense, their values are incoherent – not thought through. The desires they express at any particular time are those tapped by the particular question posed. That question may evoke a central concern or a peripheral one; it may help clarify the respondent's opinion or irreversibly shape it; it may even create an opinion where none existed before.

Listing a few specific effects may indicate the power an elicitor may deliberately, or inadvertently, wield in shaping expressed preferences. The desirability of possible outcomes is often evaluated in relation to some reference point. That point could be one's current (asset) position, or an expected level of wealth (what someone with my talents should be worth at time t), or that possessed by another person. Shifts in reference point are fairly easily effected and can lead to appreciable shifts in judged desirability, even to reversals in the order of preference. Consider, for example, how one might think about the same safety programme conceptualized in terms of lives saved or lives lost, with the respective reference points of the current situation or an ideal one. As one gets closer to an event with mixed consequences, the aversiveness of its negative aspects may increase more rapidly than the attractiveness of its positive aspects, making it appear, on the whole, less desirable than it did from a distance. People may have opposite orders of preference for gambles when asked which they prefer (which focuses their attention on how likely they are to win) and when asked how much they would pay to play each (which highlights the amount to win). People may prefer to take a chance at losing a large sum of money rather than absorb a smaller sure loss, but change their mind when the sure loss is called an insurance premium. A relatively unimportant attribute may become the decisive factor in choosing between a set of options if they are presented in such a way that that attribute affords the easiest comparison between them.

Three important features of these shifting judgements are:

1) people are typically unaware of the potency of such shifts in their perspective;

2) they often have no guidelines as to which perspective is the appropriate one; and

3) even when there are guidelines, people may not want to give up their own inconsistency, creating an impasse.

Limits to Setting Limits

The bottom line of a cost-benefit analysis is the analyst's best guess at the relative preponderance of costs or benefits. Before action can be taken, one must know how good that best guess is. Depending upon the breadth of the confidence intervals on the "best-guess" cost-benefit ratio, one might want to collect more data, install back-up systems to reduce some of the uncertainties, or abandon the project for one whose consequences are better known.

The analysts' standard practice for acknowledging and accomodating uncertainty in their inputs is through the judicious use of sensitivity analyses. The final calculations are repeated, each time using an alternative value of one troublesome probability or utility. If each reanalysis produces similar results, then the case is made that these particular errors do not matter. One way of viewing the research on judgemental biases described in the previous section is that it merely points to additional sources of error calling for sensitivity analysis.

Unfortunately, however, there are no firm guidelines as to which inputs might be in error or what is the appropriate range of possible values to be tested. The possibility of judgemental biases would, for example, be considered only if the analyst were aware of the relevant research and took it seriously. A further problem with sensitivity analysis is that it typically tells us little about how the uncertainty from different sources of error is compounded, or about what happens when different inputs are subject to a common bias. The untested assumption is that errors in different inputs will cancel one another out, rather than compound in some pernicious way (Fischhoff, 1977).

The reasonableness of such an independence assumption seems weak when a set of judgements is elicited with the same procedure, inducing the same perspective. For example, asking about preferences in a mode that incorporates a reference to dollar values might persistently deflate the expressed importance of environmental or other less tangible values. To take an example from the elicitation of judgements of fact, the U.S. Reactor Safety Study called upon its experts to assess unknown failure rates by the "extreme fractiles" method, choosing one number so extreme that there was only a 5 per cent chance of the true failure rate being higher and a second number so low that there was only a 5 per cent chance of the true rate being lower. Research conducted with a variety of other tasks and subjects has shown that this technique routinely produces too narrow confidence intervals, so that the precision of these estimates is systematically exaggerated (Lichtenstein et al., 1977).

Even if sensitivity analysis could handle the compounding of uncertainty, in some contexts it completely misses the point. Many of the effects discussed under the rubric of the lability of values reflect the introduction of new, possibly foreign, possibly distorted perspectives into a decision-making process. Invocation of sensitivity analysis will not excuse the imposition of an elecitor's perspective on the respondent. Nor will it handle shifts in perspective that lead to reversals of preference.

In the end, determining the quality of an analysis is a matter of judgement. Someone must intuit which inputs are dubious and which alternative values should be incorporated in sensitivity analyses. Essentially, that someone must decide how good his or her own best judgement is. Unfortunately an extensive body of research suggests that people are overconfident in the quality of their own judgement. Indeed, people have been found to be so overconfident in their degree of general knowledge that they will accept highly disadvantageous bets based on their confidence judgements. Furthermore, this bias seems to be impervious to instructions, familiarity with the task, question format and various forms of exhortation toward modesty (Lichtenstein et al., 1977; Fischhoff et al., 1977).

A particularly relevant version of this overconfidence emerges when people are asked to judge the completeness of the representation of a problem. Research here has shown a persistent tendency to underestimate what is left out and overestimate what is known. As before, while most of this psychological evidence is derived from work with lay people, there is some systematic and considerable anecdotal evidence of similar processes at work with experts. Among the generic kinds of issues whose omission might not be adequately noted by energy analysts are:

1) the imaginative ways in which human error can mess up a system;
2) the range and rate of possible changes in people's values and behaviour regarding energy consumption;
3) the number of unknown or undetected physical, biological, or psychological effects of a new system; and
4) the interrelation between system components (e.g., common mode failures or the possibility of a system failing because a back-up component has been removed for routine maintenance (Fischhoff *et al.*, 1978).

No analysis, performed in real time and with finite resources, claims to be complete or error free. Indeed, all responsible analysts include sensitivity analyses with their reports. The preceding discussion suggests, however, that it is hard to assess the adequacy of these analyses. We have an urgent need for a better understanding of what errors may enter into an analysis, how virulent they are, how they are propagated and compounded through the analysis, what can be done to reduce their impact, how we can assess their total impact, and what that assessment means in terms of action. In a sense, what we need is an error theory for cost-benefit analysis, supplemented by some empirical study of the fallibility of analyses conducted in the past.

The qualifications accompanying many (or most) analyses include reference to what could have been done with greater time and resources. These two commodities are, however, always going to be limited and we must know how well cost-benefit analysis serves us under realistic constraints. One conclusion of such an assessment might be that cost-benefit analysis is useless unless X per cent of the total budget can be invested in it; another might be that virtually all the value of a cost-benefit analysis comes from structuring the problem and conducting a few back-of-the-envelope calculations within that structure. A third possible conclusion is that, in most situations, the judgemental components of cost-benefit analysis are so essential and so deeply buried that conducting a formal analysis merely creates an aura of solvability around problems that are quite dimly understood.

Limits of Scope

Like all other procedures, cost-benefit analysis deals with only a segment of the management problem. The crucial question here is whether that segment can stand alone and is able to contribute to the rest of the process or whether its internal logic disintegrates when confronted with broader realities.

The segment addressed by cost-benefit analysis is that most amenable to formal analysis and least accessible to individuals without technical expertise. This is most certainly true in variants that rely heavily on tools like shadow pricing or revealed preferences to deduce what the people want without asking them. Some variants, like decision analysis, try to overcome this bias by incorporating elicitation procedures that can, in principle, be used with corporate executives, government regulators or people off the street. Despite such efforts, however, the very sophistication and centralization of the analysis gives added weight to the opinions of those who are articulate and close to the analyst.

When analytic resources are limited, the analyst must take cues from someone about how to restrict the alternatives and consequences considered. That someone is likely to be the one who commissioned the study. If commissioners all come from one sector of society and consistently prefer (or reject out of hand) particular kinds of solutions or consequences, a persistent bias may be produced. Such bias would determine what issues are never analysed and how results are presented. If the commissioners are public officials, there may be a strong predisposition toward reports that bury uncertainties and delicate assumptions in sophisticated technical machinations or in masses of undigested data.

If one examines the public criticisms to which cost-benefit analyses are subjected, a number of themes emerge. One is that there is usually a population of experts who are angered because the sensitivity analysis did not include what they believe to be appropriate alternative values for some inputs. When these experts view themselves as having qualifications rivalling those of those experts who conducted the study, the obvious implication is that even the elite community involved is somewhat restricted, perhaps for reasons of political or academic power.

A second group of critics views the segmentation of participants as a more serious issue than the substantive topic at hand. Their main concern is that the use of analysis transfers power for societal decision-making to a technical elite, in effect disenfranchising the lay citizenry.

One might argue that given the vagaries of lay judgement described above, such a transfer of power is in the best interests of even that lay public. "Let someone competent do the job; we'll all be better off." The counter-argument has several facets. One is that every analysis requires a variety of judgements that might just as well be performed by lay people. Regarding questions of fact, when they are forced to go beyond their tools and data and rely on intuition, experts may be little better than non-experts. Regarding questions of value, being close to the action should not confer superiority on experts' beliefs. The second part of the argument is that there are higher goals than maximizing the efficiency of a

particular project. These include developing an informed citizenry
and preserving democratic institutions. The process may be more
important than the product, making it important to devote the
resources needed to make meaningful public participation possible.

Such participation requires new tools for communicating with the
public, both for presenting technical issues to lay people and for
eliciting values from them. It may also require new social and legal
forms, such as hiring representative citizens to participate in the
analytical process, thereby acquiring the expertise needed to confer
the informed consent of the governed on whatever decision is
eventually reached. Such a format might be considered a science-
court with a lay jury. It would consider cost-benefit analysis as one
input to its proceedings. It would also place the logic of
jurisprudence above the logic of economic analysis, acknowledging
that there is no formal way to summarize the issues at hand.

A third group of critics objects to the separation of the energy
problem from the broader context of social issues. The critics often
fight dirty or irrationally (from the perspective of the formal analyst)
because they view the cost-benefit analysis as one arena in which
political struggles are waged. Those struggles have a different logic
than that of economic analysis. In them, it may be fair to engage in
unconstructive criticism, viciously poking holes in analyses if the
results do not support one's position. It may even be legitimate to
ridicule or chastize analysts for ignoring issues (like income distribution) that were outside their analytic mandate.

Some proponents of this position would argue that the very
reasonableness of formal analysis involves a political-ideological
assumption, namely, that society is sufficiently cohesive and
common-goaled that its problems can be resolved by reason and
without struggle. Although such a "get on with business" orientation
will be pleasing to many, it will not satisfy all. For those who do not
believe that society is in a fine-tuning stage, a technique that fails to
mobilize public consciousness and involvement has little to
recommend it.

Thus, there are logics other than that of cost-benefit analysis,
coming from legal, political, even revolutionary theory. Like the
various ways of implementing the basic cost-benefit framework, each
embodies both ideological predispositions and notions of how society
operates. Considering these perspectives and the impact that
problem formulation can have on people's judgements of their own
values and the tendency for such analysis to create an aura of
solvability, cost-benefit analysis no longer appears as a value-neutral
procedure. This does not mean that it is not or cannot be made into
the one technique most compatible with or capable of incorporating
the broadest range of values in a particular society. It does mean
that a political position of sorts is being taken when one adopts the
procedure.

Like most choices involving ideologies, questions of taste are somewhat disciplined by questions of reality. Those who oppose cost-benefit analysis are responding in part at least to social concerns and facts to which the analysis is relatively or totally deaf. Attention to these concerns can strengthen an analysis and heighten its impact. For example, an analysis that ignores questions of equity will often be overturned by those who come out on the short-end of the project in question. Rather than let their work become a numbers game with no real effect, many analysts have attempted to exploit the kernel of truth in their critics' arguments and incorporate equity considerations. A more political perspective might also help one realize that formal analyses deal with ideal types often having no representation in reality. It is fairly easy to become enamoured of abstractions and analyse projects that are never implemented in the way or at the time they are proposed (Majone, 1975). Although the method may be tricky, one could respond to this challenge by considering ensembles of possible representations of the proposed project (i.e. ways in which it might be carried out) or by requiring periodic updates of an analysis as the facts change. A broader perspective could motivate analysts to specify the assumptions about society upon which their analyses are predicated and heighten their sensitivity to the tenuousness of those assumptions. In the extreme, it might even lead them to reject analytical mandates that separate projects from their social context in ways that are not meaningful.

Conclusion

How well does cost-benefit analysis fare according to the various criteria listed in Table 1? What work is needed to make it fare better? Even though the preceding discussion has focused on some of these topics at the expense of others, a few words about each will give a flavour of what a fuller consideration might reveal.

Logically Sound

The cost-benefit family of analytic procedures has a logical foundation that is both carefully thought out and widely accepted in both political and academic circles. As a result, both its failings and its assets are better documented than those of its competitors. Although there are still important technical and conceptual problems to work out within the cost-benefit framework (Pearce, this volume), a useful investment of energy might be trying to clarify the relationship between that framework and the logic (or ostensible logic) of other approaches. Are they really incompatible? Can cost-benefit analysis be elaborated to incorporate the elements of truth embodied by the alternatives? Perhaps the weakest competition is provided by the "logic" uncovered in studies of people's intuitive decision making. Yet, even here, it is worth asking whether there is not a method in people's apparent madness. Are there not decision-making criteria

overlooked by formal analysis yet essential for human welfare or
psychological well being?

Implementability

Like other computational enterprises, cost-benefit analysis stands or
falls on the strength of its inputs. While enormous strides have been
made to develop a cumulative data base on various topics, all too
often the analyst is forced to rely on intuitive judgements.
Judgements of fact tend to be subject to persistent biases that are
only now beginning to be understood. Further work is needed here,
particularly in studying the judgements of experts. Judgements of
value tend to be highly labile and subject to complex and subtle
manipulation by the questioning procedure used. Research is needed
to produce techniques and settings that enable respondents to
elucidate their own opinions.

Politically Acceptable

In contrast to the political objections to cost-benefit analysis raised
earlier, one may cite a number of fundamental assets. The most
important of these is the explicit nature of its structure and inputs,
all of which are in principle open to examination and revision. To
realize this potential advantage, several developments are needed.
One is procedures for communicating technical issues to lay people
(including regulators and legislators without a technical background)
so that they can offer reasoned critiques. A second need is to
develop some way for critics to perform their own sensitivity
analyses, incorporating their own alternative values for various
inputs. Such an opportunity might produce some surprising results,
showing the conclusions of analyses to be much more (or less) robust
than they initially appeared. It could also help allay fears that these
conclusions represent the result of ingenious number-fudging by the
analysts. In the back of many cynical critics' minds must lurk the
thought that the experts have played around until they found a
constellation of values that looks benign but produces the
recommendation most favourable to a particular point of view.

Respects Institutional Constraints

The initial popularity of cost-benefit analysis probably was due to its
fitting well into the way in which business was done in various seats
of government. Its continued success may be due to the ability of
proponents to shape legal and governmental proceedings to
accommodate this tool further. Its future prospects may depend on
the successful resolution of several persistent problems. One is that
it assumes a single decision-maker; difficulties arise when there are
many hands involved and many views to be incorporated. A second
problem is that it is a one-time analysis; as a result, it is not as

responsive to changing contingencies, preferences and scientific data as are the bureaucracies it is designed to serve. A third problem is that it requires a level of analytical expertise not possessed by many of the individuals involved in its use, producing aberrations and frustrations.

Open to Evaluation

As mentioned, cost-benefit analysis claims to be, if nothing else, open to inspection. To realize the promise of this claim, several developments are needed. Psychological research must find ways to help people appraise the limits of their own knowledge or, failing that, ways to assess how confident they should be, given how confident they say they are. Theoretical efforts are needed so that analyses produce better assessments of their own limits and derive the action implications of that cumulative uncertainty. Since such efforts have their own inherent limits, empirical work is needed to review past analyses to explore their foibles and contribution (or lack of it) to the management process.

Creates No Side-Effect

Disenfranchising the lay public is one possible side effect of the wide-scale adoption of cost-benefit analysis about which some segments of the public are quite agitated. Denigrating the importance of consequences that cannot readily be expressed in dollar or other quantitative terms (e.g., extinction, aesthetic degradation) is another. Making a fetish out of currently enjoyed benefits is a third (Mishan, 1974). In general, though, the implications of having a cost-benefit society are poorly understood.

Promotes Long-Term Effective Management

Cost-benefit analysis, particularly its sensitivity analysis component, has been instrumental in setting the research agendas of those concerned with understanding the effects of technological projects. To the extent that the priorities of such analyses are correct, this is a major contribution to creating a base of data relevant to sound management. It has also provided a framework within which talented economists could apply themselves to these problems. Finally, although it is criticized for emphasizing product over process, the cost-benefit analysis framework has probably raised the level of debate in many settings and broadened the understanding of even its harshest critics.

EPILOGUE

There is no verdict on cost-benefit analysis *per se*. One must consider it in the light of alternative approaches and in the context of particular situations that might accentuate its strengths or weaknesses. One must consider not only cost-benefit as it is today, but as it can be improved. One must consider not only the nice idea and the sparkling theory, but the integrity with which it will be applied.

Given the limits to human judgement and consensus described here, it is unlikely that cost-benefit analysis in a pure form will ever be practised or followed anywhere. The critical question then becomes, does it degrade gracefully?

Acknowledgement

This research was supported by the Advanced Research Projects Agency of the Department of Defense and was monitored by Office of Naval Research under Contract N00014-79-C-0029 (ARPA Order No. 3668) to Perceptronics, Inc.

References

Fischhoff, B. (1975). Hindsight $\neq$ foresight: the effect of outcome knowledge on judgment under uncertainty. *Journal of Experimental Psychology: Human Perception and Performance* 1, 288-299.

Fischhoff, B. (1977). Cost benefit analysis and the art of motorcycle maintenance. *Policy Sciences* 8, 177-202.

Fischhoff, B. (in press). Decision analysis: clinical art or clincial science? In "Proceedings of the Sixth Research Conference on Subjective Probability, Utility and Decision-Making", Warsaw, 1977. (L. Sjoberg and J. Wise, eds.).

Fischhoff, B., Slovic, P. and Lichtenstein, S. (1977). Knowing with certainty: the appropriateness of extreme confidence. *Journal of Experimental Psychology: Human Perception and Performance* 3, 342-355.

Fischhoff, B., Slovic, P. and Lichtenstein, S. (1978). Fault trees: sensitivity of estimated failure probabilities to problem representation *Journal of Experimental Psychology: Human Perception and Performance* 4, 342-355.

Fischhoff, B., Slovic, P. and Lichtenstein, S. (in press). Knowing what you want: measuring labile values. In "Cognitive Processes in Choice and Decision Behaviour", (T. Wallsten, ed.). Erlbaum, Hillsdale, New Jersey.

Harriss, R., Hohenemser, C. and Kates, R. (1978). Our hazardous environment. *Environment* 20, 6-15 & 38-41.

Keeney, R.L. and Raiffa, H. (1976). "Decisions with Multiple Objectives". Wiley, New York.

Layard, R. (ed.) (1974). "Cost Benefit Analysis". Penguin, New York.

Lichtenstein, S., Fischhoff, B. and Phillips, L.D. (1977). Calibration of probabilities: the state of the art. In "Decision-Making and Change in Human Affairs" (H. Jungermann and G. de Zeeuw, eds.). D. Reidel, Amsterdam.

Majone, G. (1975). The feasibility of social policies. *Policy Sciences* **6**, 49-69.

Mishan, E.J. (1974). What is wrong with Roskill? *In* "Cost-Benfit Analysis" (R. Layard, ed.). Penguin, New York.

Raiffa, H. (1968). "Decision Analysis". Addison Wesley, (Reading), Mass.

Ross, L. (1977). The intuitive psychologist and his shortcomings. *In* "Advances in Social Psychology" (L. Berkowitz, ed.). Academic Press, New York.

Slovic, P., Fischhoff, B. and Lichtenstein, S. (1977). Behavioral decision theory. *Annual Review of Psychology* **28**, 1-39.

Stokey, E. and Zeckhauser, R. (1978). "A Primer for Policy Analysis". Norton, New York.

Tversky, A. and Kahneman, D. (1974). Judgement under uncertainty: heuristics and biases. *Science* **185**, 1124-1131.

THE ROLE OF MODELLING: A MEANS FOR ENERGY/ENVIRONMENT ANALYSIS

Robin L. Dennis

*National Center for Atmospheric Research,
Boulder, Colorado, U.S.A.*

This paper deals solely with modelling and analysis applied to long-term management and strategy-analysis. It is believed, however, that the experiences and conclusions discussed here are more general.

The need for policy analysis is stated and the proper role of modelling for policy analysis is defined. The model's role as a means for analysis is described, using experience from case studies. Inadequacies that hinder the fulfilment of this role in energy/environment modelling are then discussed. Finally, decision theory is presented as a possible way to compensate for these inadequacies.

Need for Analysis

There are three major reasons for energy/environment strategy or policy-analysis.

Management of Environmental Effects Must Take a Longer-term Perspective We are coming to a time when we can no longer afford to plan without concern for the future and assume that all problems that "unexpectedly" arise can be handled in a "crisis" or a "technical quick-fix" mode. This implies that we need to understand and model possible effects far into the future.

Energy/Environment Questions are Multi-Media in Character Environmental impacts from energy affect the media of land, air, water and social systems. The impacts are also widely distributed in time and space. This implies a great need to collect and organize information.

Energy/Environment Questions are Multi-objective in Character Any long-range planning has embedded in it a mixture of social, environmental, energy-related and other objectives. Some trade-offs between objectives will probably be necessary because the different objectives are likely to be in partial conflict. Recognizing and dealing with multiple objectives is becoming important in energy strategy-analysis. This implies a need for a consistent focus and an even-handed evaluation of alternative strategies.

Modelling is the primary tool available to do this policy analysis, and in many cases the only tool we have for long term policy analyses.

Role of Models

For policy modelling, the model serves as a vehicle for the analysis and is not an end in itself. The model should help focus and clarify the analysis, not be the analysis. This is simple to say, but difficult to do. Systems analysts are concerned with this supportive role of the model for policy analysis, but usually modellers are seldom, if ever, concerned with building and using models in this manner.

Type of Modelling

The use of modelling in the policy analysis must be dynamic and evolving, and should emphasize the process as much as the model. The act of modelling may be more important than the resultant model. The models should therefore be as simple and transparent as possible.

The model must also be capable of supporting a wide-ranging analysis. The analysis should be able to move beyond our own biases and push beyond accepted "common sense". Thus in this context simulation-analysis is one of the most useful types of model.

Optimization models are more cumbersome for the wide-ranging analysis required, and are therefore less useful than simulation for long-range policy analysis. Optimization analysis can be used together with simulation analysis, but it should be used with a maximum of sensitivity and constraint-analysis. This paper will discuss only simulation analysis experience.

Energy/Environment Case Study Experience

Four energy/environment case studies have been performed at Wisconsin and IIASA[1] by an interdisciplinary team of researchers (Foell, in press; Foell et al., 1978a). The four case studies were for the regions of Austria, Rhônes-Alpes, France, a district of the German Democratic Republic and the State of Wisconsin, U.S.A. The work centred around simulation analysis and used a family of models to carry out the simulations (Foell et al., 1978b). Scenarios were used to bind the family of models together for a consistent analysis and to bring together the technical and non-technical elements of the analysis.

---

[1]International Institute for Applied Systems Analysis, (Laxenburg, Austria).

One of the important lessons of this policy-analysis work is the understanding that the proper role of the models and modelling is to provide a means, i.e. focus and vehicle, for the policy analysis. This conclusion as to the role of the model is not new and is similar to conclusions arrived at in ecological modelling (Holling, 1978); early systems analysis experience (Quade and Boucher, 1968); and resource management modelling (Fiering, 1976).

Three major elements comprise the case study modelling experience mentioned above:

1) Modelling makes assumptions explicit.
2) Modelling promotes systematic exploration of alternatives.
3) The models will always be incomplete.

There are two additional aspects of the modelling analysis experience that are relevant to a discussion of the proper role of models:

1) Modelling gives a reservoir of expertise.
2) For policy analysis, the modelling should not be a one-time affair.

Three Major Elements

Assumptions Explicit Making assumptions explicit helps focus the discussion and clarify the issues. In the energy/environment case studies, detailed, disaggregated trends of electricity-demand growth for the different economic sectors were developed. In the Wisconsin and the Austrian case studies, when these demands were aggregated to a total electricity-demand growth, the growths projected were consistently lower than those projected by the electricity companies. Since the disaggregated components of the electricity demand were detailed in the model and it was shown that no major part had been omitted, the higher growth rates projected by the electricity companies could not be explained. With the assumptions explicitly available in the model, the argument of need for electrical energy could be addressed very clearly. For Austria, such explicitness could be important because it indicated that there may well not be the urgency to make a quick decision concerning nuclear power as suggested by proponents since they were using what appeared to be relatively high electricity demand projections.

Systematic Exploration In the Austrian case study, the modelling analysis allowed a systematic examination of several possible future SO_2 regulations as to their effect on reducing SO_2 emissions and their effectiveness in reducing the health-impacts of air-pollution. The model facilitated a comparison not only between SO_2 regulations, but also a comparison between energy conservation and the SO_2 regulations. For Vienna, the unexpected conclusion of the analysis was that energy conservation would be expected to contribute more to a reduction of public-health impact due to air pollution than the assumed air pollution regulations themselves. These aspects are now being considered in Vienna's future energy planning. This unanticipated result points out the role of the model in the analysis.

Always Incomplete Environmental analyses will always be incomplete. For example, in calculating air pollution impacts, many analyses only include the energy supply sector, leaving out the potentially very important end-use sectors. No analysis has yet included both long-range and short-range SO_2 transport, yet we know both are important. Even if all ranges of pollutant transport could be and were included in the analysis, only a few air pollution health-damage and biota-damage functions exist, and they have a large degree of uncertainty associated with them. Information about impacts and potential impacts is continually increasing, but the information-base is never complete enough to quantify everything that is considered important for the analysis. Thus, in almost all cases, we are dealing not with environmental impacts, but with environmental-impact indicators. This is an important limitation to the direct use of model numbers and increases the importance of providing an analysis of those numbers.

Additional Elements

Reservoir of Expertise The process of building a model and performing analyses with it develops an expertise that may be applied to problems not related to the ones for which the model was built. In Wisconsin, as a result of the building of a model to calculate the energy demand of the commercial sector, one of the researchers was asked to participate on the committee that proposed the first Wisconsin insulation standards for commercial buildings. Since general building heat-loss had already been quantified on an engineering basis in the model, the model was capable of quick adaption to provide some insights on the new question of insulation standards.

Not a One-time Affair As time goes by, unanticipated changes happen; new options become available, new information on old options becomes available or the suspected causal agent of an impact changes. As a result, the model calculations can quickly become out-dated. The analysis, however, may not be so quickly out-dated where the emphasis is on changes and differences, rather than absolute values. Nonetheless, old analyses are viewed as suspect when policy makers know parts of the model are out-of-date. The analysis is a continual process, as indeed the questions also change with time.

The Wisconsin research group has updated the scenarios for Wisconsin several times in the past five years. Significant changes have evolved and energy-demand growth that several years ago was projected as being associated with conservation measures and patterns is now associated with moderate and higher energy demand growth (M. Hanson, University of Wisconsin, private communication). Differences between old and new scenario sets are as informative for a long-term analysis as the analysis of differences within a scenario set. The State of Wisconsin is also asking the Wisconsin research group for new analyses that older scenario sets were not designed to answer. The new analyses use the same model base, but focus on different parts of the model or incorporate new information into the models. It is this adaptive process in the analysis, for which the model is the vehicle, that is important for good policy analysis.

One important point bears repeating. The models are always incomplete. More questions will be asked of the analysis than can be quantified for the analysis by the model. Nevertheless, the model serves its role as a means of analysis in several ways: focusing the analysis and discussion of issues, clarifying the assumptions, pointing out the well-known and the poorly-known, and facilitating discussion of differences and uncertainties that have relevance to the policy issues.

Major Inadequacies

Many criticisms of energy/environment modelling inadequacy are really complaints about lack of information. Either data is available and a model has simply not yet been formulated and built, or the scientific basis for formulating a model is not yet sufficient or available. These are perpetual complaints because they reflect the constant incompleteness of the models used for the analysis.

There exists, however, a more fundamental inadequacy in energy-/environment analysis that truly hinders the use of the model as a vehicle for policy analysis. The inadequacy can be expressed as two seemingly separate parts:

1) Environmental analysis is static and not on an equal basis with the energy analysis, and
2) multiple objectives are not well adddressed.

Static and Unequal Analysis

Much of the energy-related environmental analysis has the static character of an environmental impact statement. The energy plan is developed and then an environmental impact analysis proceeds to list a set of impact indicators. The environmental analysis occurs after the fact of the energy plan as a final required check to demonstrate some level of acceptability. The environmental analysis and the energy analysis rarely occur concurrently, and with very few feedbacks.

The environmental analysis should be playing a more equal and active role in the development of the energy strategies. The fact that it does not, partly reflects the usual institutional division of decision-making for energy matters and environmental matters. The case study work at Wisconsin and IIASA attempted to integrate environmental policies with the energy policies via the consistent formulation of scenarios. Whilst this is a significant improvement, the results are not truly satisfactory because the environmental impact indicators are still passively connected to the energy analysis. It remained difficult to develop a combined analysis in which there is an equal interplay between energy and environmental objectives.

Optimization models actually appear to be better in this respect in that environmental objectives and constraints can be formally included in the analysis (E. Cherniavsky, Brookhaven National Laboratory, personal communication). It is felt, however, that the measure of environmental impacts that have been used in optimization models up to now, are too far removed from the real impacts to provide any truly useful environmental analysis. This appears to be as much due to the computing constraints imposed on optimization-type models as due to lack of model formulations that would allow the inclusion of more realistic impact indicators.

Multiple Objectives Not Well Addressed

Much of environmental analysis stops, as do impact statements, with a list of impact indicators, leaving any evaluation and trade-off between, for example, environmental, economic and energy consider-ations to a completely informal process. Formal means of weighting and combining impact indicators of disparate character, such as nuclear proliferation risks, SO_2 air-pollution impact on public health, and damage to ecosystems from oil spills, are seldom used. Formal means of addressing multiple objectives, such as environmental impacts, energy system costs and perceived benefits from energy, representing different objectives of a single decision-maker or of different interest groups are left to informal "mental models" and to subjective discussion.

Yet, even though the models are incomplete, it has been shown that clarifying and quantifying assumptions and doing a systematic analysis with them provides much more information than is provided by an unstructed analysis. This is not only the case for description of physical processes, but is also the case for cognitive activity (Hammond, 1978a).

Possible New Directions

One area of analysis appears to have the potential to significantly contribute to the alleviation of the two inadequacies described above. This area of analysis is variously called human judgement theory, decision analysis and multi-attribute utility theory (Kaplan and Schwartz, 1975; Hammond, 1978b; Keeney and Raiffa 1976). Because we realized the inadequacies above, the Wisconsin and IIASA energy/environment case studies included some preliminary use of

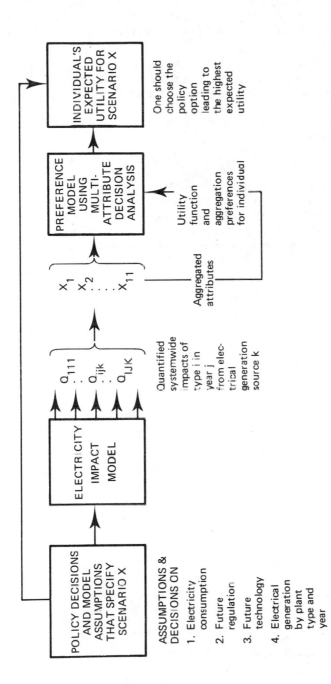

Fig. 1 Example of a composite environmental impact model. For full details, see Buehring *et al.,* 1976.

the multi-attribute utility theory to aid the overall analysis (Buehring *et al.*, 1976; Keeney, 1975; Buehring, 1975). It was perceived that the multi-attribute utility theory could provide four improvements to the analysis:

1) formally define preferences for disparate indicators;
2) formally weight and combine disparate indicators;
3) formally include uncertainties in the evaluation of preferences, and
4) provide a more formalized link of environmental impacts back to the scenario assumptions.

This leads to the concept of a composite environmental impact model as shown in Figure 1.

The preference structure of decision-makers and energy-policy analysts close to decision makers of Wisconsin, Rhônes-Alpes, France and the German Democratic Republic were evaluated and the use of the multi-attribute utility analysis technique to evaluate and rank preferences for scenario outcomes was demonstrated (Buehring *et al.*, 1976). Figure 2 shows the preferences (utility functions) of two

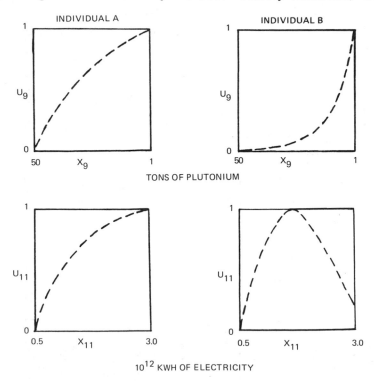

Fig. 2 Selected conditional utility functions for two individuals. For full details, see Buehring, 1975.

individuals for two different scenario output indicators (attributes) — electricity generated and plutonium wastes generated.

Figure 2 clearly shows that very different preferences can be exhibited for each indicator. For example, the two individuals have very different views about the growth of energy. Individual A will always prefer more electricity in the range given, although the utility of each new increment of electricity is decreasing. Individual B, on the other hand, displays a preference that implies there is an "optimal" level of electricity use in the given range. The perceived disbenefits of too little or too much electricity compared to the desirable amount are considered to outweigh any perceived benefits of the use of the electricity.

What We Learned

In using the multi-attribute utility analysis it became clear that one of the most important benefits of this analysis, if not the most, was the process itself. The process benefited both the modeller and the decision maker. First, in defining the aggregated attributes, X_i, with the policy maker (see Figure 1) the analyst/modeller received the message from the future user of the information as to what were considered to be the relevant indicators to adequately evaluate the issues. The modeller may leave something out through ignorance or nervousness about the uncertainties involved in making a quantitative calculation. This would then have to be reconciled in some way, for example, by a surrogate indicator, for the analysis to be meaningful to the policy maker. Second, in defining her/his preference structure, the decision-maker comes away with a much more sophisticated and precise understanding of personal preferences, regardless of whether or not the resulting utility function is ever used. At the present stage of energy/environment analysis, these are the most important benefits of the analysis.

The process of defining preference structures can also clarify fundamental issues and point out fundamental conflicts of objectives. This has the same function as making the assumptions explicit for the physical models. An example, in the author's mind, is some of the work on nuclear power risks by the joint IIASA/IAEA risk group (Otway and Fishbein, 1977; Bowman et al., 1978). One can interpret these findings in the following way. The researchers were looking for the attributes that would provide a measuring set for the determination of preference structures concerning nuclear power. They found a common set of indicators that was sufficient to define the problem and was based on available information. This would tell the modeller that the indicators and information generated by the technical models did not have serious gaps.

The important point, however, is that with the same set of values for the indicators (analagous to evaluating the same scenario at a given point in time) people could reach opposite conclusions, i.e.

whether they were for or against nuclear power. It was the cognitive process of evaluating, weighting and the combining the indicators that was producing the conflict between the *pro* and *con* people. At the present time, adding reams of technical information with more sophisticated models will not alter the situation. In this example, if a decision were required now, the decision analysis process pointed out the area of conflict, i.e. nuclear risks. Given that one of the positions against nuclear power appears to centre on a substantial skepticism of the engineering estimates of nuclear risks, a double-bind is created, because a better knowledge of these nuclear risks may only come about via actual experience. The decision-analysis process has pin-pointed the essence of the conflict rather precisely. This is important for relevant decision-making.

Failures

Of the four advantages perceived with the multi-attribute decision theory approach (listed above), the last two were failures at this stage of the energy/environment experience. Notwithstanding the difficulty of quantifying uncertainty for the descriptive scenario-generating models, the task of coping with the definition of preferences was so enormous and time consuming that the question of uncertainty could not be approached directly. Certainly, preferences may be influenced by perceptions of uncertainty, but this remained buried within the preference analysis.

The time-consuming character of defining preference structures also meant that the composite environmental impact model of Figure 1 was never achieved. The process, although extremely valuable, was too cumbersome to connect with the scenario/simulation generation in any systematic fashion. As a result, both the environmental analysis and the decision analysis remained static in character and essentially passive, not dynamic.

Decision-makers showed great reluctance to go beyond the first preference structuring phase because of the large time commitment required. This effectively limits the potential of decision analysis to the value of its process. In situations where the decision-makers were not at all used to simulation modelling and formal modelling analysis there was a great reluctance even to participate in the process of preference structuring. The cumbersome nature of the decision analysis process appears to severely limit any sort of routine use of it with scenario-simulation generation and with adaptive-management practices since they require many iterations of the models and reformulations of the scenarios.

Nonetheless, decision analysis and judgement modelling is still considered to have a potential to meet some important analysis needs for energy/environment modelling. The Symmetrical Linkage approach proposed by K.R. Hammond (1977a,b), is considered to be a promising advance from our past work with multi-attribute utility analysis. This approach has two additional aspects that appear to be very promising. First, the evaluation of preferences is interactively computerized for more rapid and easier inter-action with the policy person. Second, the Symmetrical Linkage concept starts with the environmental and preference analysis on the same footing as the more technical energy and economic analysis. The physical system models and the cognitive system models proceed together in development and analysis. A description of the linkage concept and an example of its application is given by Hammond, (1977a). These two additional features to the preference analysis may go a long way towards alleviating the deficiences in energy/environment analysis discussed above.

Combining Hammond's work with our own scenario/simulation work is a promising new direction of research. This is now being actively pursued. If this synthesis works, the role of modelling for long-term policy analysis can be greatly enriched. Some work in this direction has already been attempted by resource management analysts and the results indicate it will not be a simple nor easy task (R. Hilborn, University of Vancouver, personal communication).

As before, the role of the modelling remains unchanged - the modelling is a tool for the analysis, the search for preference functions is a tool for the analysis. Both are decision-aids. To do more would be presumptuous. The goal is better policy analysis and decision-making, not a fallacious replacement of decision-making.

Acknowledgements

The work reported here was performed at the University of Wisconsin and at IIASA. The author would like to thank Wes Foell of the University of Wisconsin for his inspiration and acknowledge his leadership in the energy/environment case studies. The author would also like to thank Bill Clark, Ray Hilborn and Buzz Holling of U.B.C., Vancouver, Canada for their valuable comments, discussion and sharing of experience.

References

Bowman, C.H., Fishbein, M., Otway, K.J. and Thomas, K. (1978). "The Prediction of Voting Behaviour in a Nuclear Energy Referendum" (report). IIASA Research Report, Laxenburg, Austria.

Buehring, W.A. (1975). "A Model of Environmental Impacts from Electrical Generation in Wisconsin" (Ph.D. thesis). Department of Nuclear Engineering, University of Wisconsin, Madison, Wisconsin.

Buehring, W.A., Foell, W.K. and Keeney, R.L. (1976). "Energy/ Environment Management: Application of Decision Analysis" (report). IIASA, Laxenburg, Austria.

Fiering, M.B. (1976). *Natural Resources Journal* 16, 759.

Foell, W.K. *et al.*, (1978a). "IIASA Austrian Regional Energy/ Environment Study" (report). IIASA, Laxenburg, Austria.

Foell, W.K. *et al.*, (1978b). "A Family of Models for Regional Energy/Environment Analysis" (report). IIASA, Laxenburg, Austria.

Foell, W.K., ed. (1979). "Management of Energy/Environment Systems: Methodologies and Case Studies". John Wiley and Sons, New York, (in press).

Hammond, K.R. (1978a). Toward increasing competence of thought in public policy formation. *In* "Judgement and Decision in Public Policy Formation, (K.R. Hammond, ed.). Westview Press, Boulder, Colorado.

Hammond, K.R. (1978b). "Judgement and Decision in Public Policy Formation". Westview Press, Boulder, Colorado.

Hammond, K.R., Klitz, J.K. and Cook, R.L. (1977a). "How Systems Analysts Can Provide More Effective Assistance to the Policy Maker (report). IIASA, Laxenburg, Austria.

Hammond, K.R., Mumpower, J.L. and Smith, T.H. (1977b). IEEE Transaction on Systems, Man and Cybernetics, No.5.

Holling, C.S. (ed.). (1978). "Adaptive Environmental Assessment and Management". John Wiley and Sons, New York.

Kaplan, M.F. and Schwartz, S. (eds.). (1975). "Human Judgement and Decision Processes". Academic Press, New York.

Keeney, R.L. (1975). "Energy Policy and Value Tradeoffs" (report). IIASA, Laxenburg, Austria.

Keeney, R.L. and Raiffa, H. (1976). "Decisions With Multiple Objectives: Preferences and Value Tradeoffs". John Wiley and Sons, New York.

Otway, H.J. and Fishbein, M. (1977). "Public Attitudes and Decision-Making" (report). IIASA, Laxenburg, Austria.

Quade, E.S. and Boucher, W.I., eds. (1968). "Systems Analysis and Policy Planning". American Elsevier Publishing Company, Inc., New York.

ADMINISTRATIVE ASPECTS OF ENVIRONMENTAL RISK ASSESSMENTS: THE CASE OF LEAD IN BRITAIN

Shelagh A. Staynes

*Monitoring and Assessment Research Centre,
Chelsea College, University of London, U.K.*

The assessment of the impacts and risks from pollutants released into the environment is an important part of pollutant control and of increasing interest to scientists and governments alike. But risk assessment of any pollution problem needs to be translated into policy directives and action. This process is strongly influenced by the existing legal, institutional and political machinery, which may limit the optimum utilization of the assessment.

This paper uses the example of lead pollution control in the U.K. to illustrate the complexity of using the output from risk assessment.

Lead pollution was chosen because this problem has been assessed by a number of countries, including the U.S.A. (U.S. Environmental Protection Agency, 1971), U.K. (U.K. Department of the Environment, 1974), France, and international organizations (including WHO, 1977). Yet, despite the considerable knowledge-base on impacts and risks from lead pollution, effective control poses a great deal of complex administrative and technical problems.

Lead is particularly well known as an additive in motor-fuels, and, in fact, this use represents the major source of atmospheric emissions of lead in industrialized countries. Other important, non-energy related sources of lead to the environment include the refining and smelting of lead and other non-ferrous metals; fabrication and chemical operations; lead alkyl manufacture; lead in products such as paints and primers; lead piping used in plumbing systems and lead in wastes which are incinerated or dumped on the land. Lead is thus present, and persistent, in varying concentrations in all media of the environment and the resultant pathways of lead to man are complex (Figure 1).

The first part of the paper addresses general aspects of environmental control in the U.K.,[1] including the nature of legislation and

---

[1]The terms United Kingdom and Britain are used synonymously. For

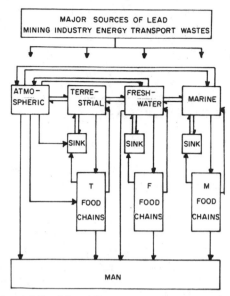

Fig. 1 Simplified model of lead in the environment (O'Brien, 1979).

institutional arrangements.   The case of lead pollution is then examined in this context.

The British Approach to Pollution Control

The British approach to pollution control is essentially the view that, while pollution is undesirable, it cannot be entirely eliminated, and (until evidence is produced to the contrary) effort should continue to be made to reduce its detrimental and nuisance effects as much as is feasible, given existing knowledge and capabilities.   At its simplest, the principle behind the control applied is that it should ensure that the adverse effect of pollution on man is the lowest practicable, having regard to the benefits and costs involved – "acceptable risk" (U.K. Department of the Environment, 1977b).

Britain adheres to the "polluter pays" principle in the sense that those responsible for pollution-generating activities bear the cost of installing and operating the necessary control equipment (such as filters in chimneys, screening or treatment plants for liquid effluents) needed to comply with the legal prohibitions.   Costs incurred in monitoring and/or research related to such controls are also (largely) borne by the polluter.   However, this principle applies in a somewhat limited sense, since it can only really work in situations where the polluter and the pollution are very clearly linked.   More sophisticated

the most part, the details and examples used in this paper refer to England and Wales only.   There are, in several cases, different arrangements for Scotland and Northern Ireland which are not described since they do not alter the principles discussed.

financial strategies with regard to pollution control are not in use, and the broader economic and social "costs" of pollution are borne by the public at large.

The keystone, however, of U.K. pollution control strategy, is "Best Practicable Means" (BPM). Britain for the most part does not use fixed standards for emissions (although certain fixed standards are applied to food and products), rather, those applying and enforcing pollution control provisions are under a duty to use BPM within the context of nationally-defined environmental quality objectives for preventing pollution. This means that authorities decide upon the stringency of the control measures specified for a particular activity or industry, having regard, among other things, to local conditions and circumstances, the current state of technical knowledge, and the financial implications, and human health and biological effects. This approach is exemplified by the work of the Alkali Inspectorate (AI), who are responsible for enforcing legal controls over atmospheric emissions of pollutants from industry. The AI operates in co-operation with industry with a considerable degree of freedom and individual discretion. Advocates of BPM claim that it provides a flexible and sensitive method of control, allowing standards to become more stringent over time. However, critics point to the resulting variability nationally with which control is enforced and the numerous "loop-holes" available to industry.

Of course, U.K. membership of the European Economic Community influences to some extent the policies adopted for pollution control, since Britain has a legal obligation to implement EEC legislation. The overall aims of the Community's Environment Policy include, *inter alia*, "the prevention, reduction and as far as possible, the elimination of pollution and nuisance", and the "maintenance of satisfactory ecological balance and the protection of the biosphere" (Commission of the European Communities, 1976). The U.K. adheres to these aims, but disagrees as to the use of uniform emission standards for their fulfilment.

A succinct summary of the British approach to pollution control was provided by the Secretary of State for the Environment, in a recent publication (U.K. Department of the Environment, 1977a):

"... One of the great advantages of our approach is that we are able to continue to respond flexibly to new pollution hazards and to maintain the pressure for more comprehensive controls of long-standing problems. Except in certain clearly defined cases, we believe that it is better to maintain gradual progress in improving the environment in the light of local circumstances and local needs than to operate through the formulation of rigid national emission standards which may in particular circumstances be either unnecessarily harsh or insufficiently restrictive.

We consider that our approach is still valid; that as in the past it enables us to secure a continuing reduction in pollution hazards; and that the greatest progress is achieved by working as far as possible in co-operation with industry . . ."

Legal Controls Over Pollution

Statute Law (Acts of Parliament) provide the main and most effective instruments for pollution control. Statutes solely concerned with pollution are few in number, since most control provisions are embodied in statutes relating to broader concerns, such as the supply and conservation of wholesome water, food of an acceptable quality, or waste disposal. Thus the range of statutes relevant to pollution control is very large and can be difficult to identify precisely. Furthermore, since the U.K. is a member of the European Economic Community (EEC) it has legal obligations to fulfil community legislation with regard to pollution control.

The nature of legal controls varies from statute to statute and subject to subject. Some outline very detailed measures for control, such as the Clean Air Act 1955 which governs control of air pollution from domestic sources and industrial furnaces. Others merely outline general prohibitions and assign powers to a particular agency to interpret and enforce them. An example of the latter case is the Alkali etc. Works Regulation Act 1906, which governs control of air pollution from chemically complex industrial processes and which is administered and enforced by the Alkali Inspectorate.

Whichever approach is adopted, a minister of the central government will always be designated as having responsibility overall, and it is largely through the powers granted to him, (e.g. to make regulations; add to lists of scheduled processes and noxious substances; hear appeals; grant exemptions), that scope is provided within statutes for flexibility to adapt legal controls to altered situations or new hazards.

Legislation has evolved from Acts for the control of emissions of particular pollutants at specific locations on a case-by-case basis (for example particular industrial works) through control on a class-by-class basis (for example, industrial processes, nationally). Further developments have been in the nature of amendments and extensions of existing legislation to cover new hazards or developments, plus the influence of EEC legislation which is of a more general nature, relating to embient quality (e.g. air quality standards, Table 1).

Institutional Arrangements

Historical development has also influenced the nature of the institutional arrangements for pollution control in the U.K. Within the central government, there is a Department of the Environment (DOE), dating from 1971, and the Secretary of State for the Environment has responsibility for the co-ordination of the pollution control activities of the numerous government departments other than DOE.

TABLE 1

*The Development and Extension Of Statutory
Controls: Atmospheric Pollution*

	EMISSIONS TO THE ATMOSPHERE OF:	LEGISLATION FOR CONTROL	DATE
CASE BY CASE	Pollutants from specific Industrial works	Alkali Act	1863
	Pollutants from chemically complex industrial processes	Alkali, etc. Works Regulation Act	1906
CLASS BY CLASS	Pollutants from domestic hearths and industrial furnaces	Clean Air Act (further provisions)	1956 1968
	Pollutants from motor vehicles	Road Traffic Acts Construction & Use Regulations	1960 1972
	Pollutants from domestic & industrial sources	Health and Safety at Work, etc. Act Control of Pollution Act	1974 1974
AMBIENT ENVIRONMENTAL QUALITY	Pollutants-general	EEC: Air Quality Standards	1970s

This multiplicity of central government departments with responsibilities for different aspects of pollution control is explained to a large extent by their functional development and organization. Departments reflect the major concerns and responsibilities of government, such as agriculture, industry, trade, energy, health and social security. Expertise for particular areas has been built up over time within the departments. Despite the occasional reorganization or creation of a new department, major departments in central government retain their strongly sectoral character. For example, the Ministry of Agriculture, Fisheries and Food (MAFF) is responsible for the control of freshwater pollution as it affects fisheries, while the main responsibility for freshwater lies with DOE. Today, with the expansion of government responsibilities to so many areas of life, departments are very large, and the range of divisions, committees and co-ordinating bodies involved, in one way or another, in pollution control is numerous and must affect the efficiency of policy formulation and control which ensues.

Further complicating the situation is the fact that the implementation and enforcement of legal controls are largely the responsibility of local authorities. In the government's view, because the effects of pollution are usually experienced first within the confines of particular localities, primary responsibility for dealing with pollution problems should rest, as far as practicable, with authorities operating at the regional or local level. Local authorities (LAs) comprise county councils and their constituent district councils, which are the semi-autonomous local government units, and various statutory authorities such as Regional Water Authorities (RWAs), established by Acts of Parliament with responsibilities for very specific concerns. In England and Wales, there are 45 county councils, 332 district councils and 10 Regional Water Authorities, all with different territorial areas, geographical, social and economic conditions. These factors are of great importance for Local Authorities formulating policies and plans, with regard to pollution control, within the guidelines laid down by central government.

Lead Pollution Control: The Implications for Risk Assessment

The characteristics of the UK control arrangements described above are evident in the detailed arrangements for the control of lead pollution in the British environment. There are legal controls and associated administrative arrangements for lead pollution (and of course, other pollutants) in: industrial, domestic and motor vehicle emissions to the atmosphere; industrial and domestic effluents released to the freshwater and marine environments; fertilizers and manures used in agriculture; food supply; consumer products and the occupational environment.

The nature of the legal instruments for control and the institutional structure of government results in a great many groups being involved in lead pollution control (Figure 2), at both central and local

INSTITUTIONAL STRUCTURE LEAD POLLUTION IN	CENTRAL GOVERNMENT	LOCAL AUTHORITIES & AGENTS	ADVISORY & RESEARCH GROUPS
ATMOSPHERE	DOE	AI, DC	CAC, WSL DHSS, MRC
FRESHWATER	DOE, MAFF	RWA	NWC, WRC MAFF, NERC
MARINE	MAFF, DT	CC	MAFF, NERC
WASTES	DOE, DI	WDA	WMAC, UKAEA
SOILS	MAFF	ADAS	ARC, NERC
FOOD	MAFF, DHSS	DC	FACC, COMA MRC, LGC
WORKPLACE	DHSS, DEm	HSE, FI	EMAS, MRC

ABBREVIATIONS USED
ADAS – Agricultural Development Advisory Service; AERE = Atmospheric Environment Research Establishment; AI – Alkali Inspectorate; ARC = Agricultural Research Council; CAC = Clean Air Council, CC = County Council; COMA = Committee on Medical Aspects of Chemicals in Food & Environment; DC = District Council; DEm = Department of Employment; DHSS = Department of Health & Social Security; DI = Department of Industry; DOE = Department of Environment; EMAS = Employment Medical Advisory Service; FI = Factory Inspectorate; HSC, HSE = Health & Safety Commission & Executive; LGC = Laboratory of the Government Chemist; MAFF = Ministry of Agriculture, Fisheries & Food; MRC = Medical Research Council, NERC = Natural Environment Research Council, NWC = National Water Council; RWA = Regional Water Authority; SCFS = Steering Committee for Food Surveillance; UKAEA = Atomic Energy Authority; WDA = Waste Disposal Authority; WMAC = Waste Management Advisory Council; WRC = Water Research Centre

Fig. 2 Main groups involved in lead pollution control

government levels. This range of groups includes not only government departments, but also specialized agents, committees and advisory bodies, government laboratories and research establishments. Some selected examples of control "systems" for different aspects of lead pollution are sufficient to demonstrate this situation and the complexity of controls in general.

Atmospheric pollution control is an interesting case. While DOE has overall responsibility for this area, different legal, institutional and administrative arrangements exist at the local level for industrial, domestic and vehicular sources of lead emissions. In the case of emissions from chemically complex industrial works (Figure 3), the Alkali Inspectorate enforce control provisions in the Alkali

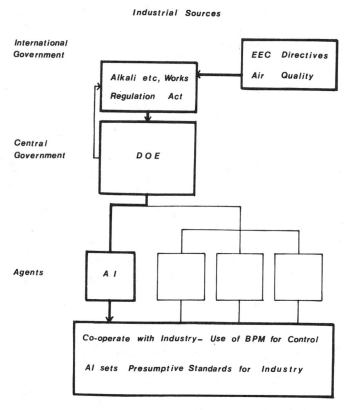

Fig. 3 Control of lead pollution in the atmosphere

Works, etc. Regulation Act, while district councils enforce provisions in the Clean Air Act for the control of emissions of smoke, SO_2 and particulate matter from industrial furnaces and domestic fires (Figure 4).

Control of the content of lead in motor vehicle emissions, however, is via voluntary agreements between industry and central government for the phased reduction of lead additives in petrol, and more generally via the Road Traffic Acts with regard to engine construction.

Lead pollution control in the freshwater environment (Figure 5) must be viewed in the context of the overall duty of Regional Water Authorities (the statutory local authorities responsible) to provide "wholesome water" and their legal responsibilities for the management of water services, conservation, supply and sewerage and sewage disposal, as well as control of freshwater pollution,

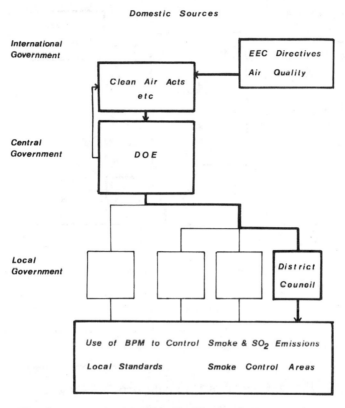

Domestic Sources

International Government

EEC Directives
Air Quality

Clean Air Acts
etc

Central Government

DOE

Local Government

District Council

Use of BPM to Control Smoke & SO_2 Emissions

Local Standards   Smoke Control Areas

Fig. 4 Control of lead pollution in the atmosphere

improvement and development of fisheries, land drainage, flood prevention, amenity and recreation of water space.

The final example is that of lead pollution of land resulting from industrial waste disposal in tips and landfill sites (Figure 6). This demonstrates the situation where a local government authority has responsibilities for pollution control as part of overall waste management in the context of its numerous other duties, such as land use planning, transport, etc.

Despite the fact that pollutants cycle through the environment and their levels and effects in one media are closely related to those in other media, pollution control is high sectoral. While account is taken (to varying extents) of the integrated nature of environmental pollution, the structure and operation of controls tend to engender the attitude that pollution in individual media is capable of being examined and treated in isolation. Hence the more frequent use of titles such as "air pollution" rather than emphasis on environmental

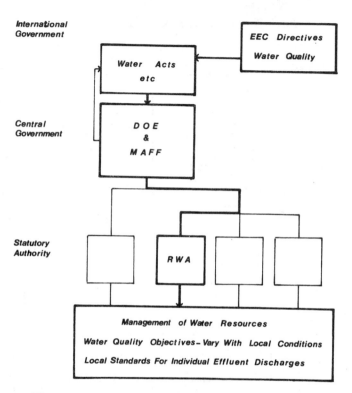

Fig. 5 Control of lead pollution in freshwater

pollution, which is readily detectable in air. Furthermore, since pollution is controlled more from the viewpoint of particular concerns such as essential resources, rather than from an overall environmental or ecological one, the degree and nature of control is closely related to the individual importance of these concerns, further engendering this sectoral quality in control arrangements. However, it must be added that scientific uncertainty as to certain environmental processes is an influential factor. The scientific literature on transfer fluxes from one environmental compartment to another is very scanty (Goodman *et al.*, 1976), and assessments tend to be orientated to the generation of "criteria", "guidelines" etc. for pollution control for individual media.

The strategies adopted for pollution control tend to be directed to those points in the cycle of a pollutant where the technological capability to effect reduced levels of contamination, or a reduction in exposure or effect can be most effectively deployed. In the U.K. for example, pollutant emissions to the air or freshwater are most easily controlled by technology applied at the point of release: the

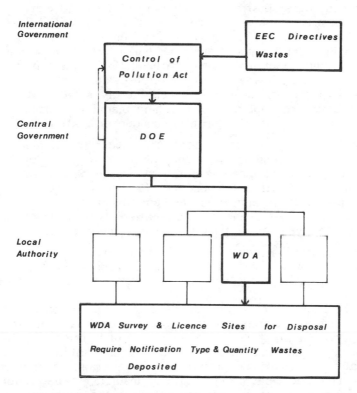

Fig. 6 Control of lead pollution on land – wastes

factory or treatment works. The numerous and diverse sources of waste, however, make it more feasible to apply pollution controls to the point of deposition. Scientific and technological effort with regard to the effects of pollution of land and pollutant transfers from the land to other media is directed to the specification of "safe" disposal sites.

Whatever the advice and guidance available from central government, it is nonetheless the local authorities who must interpret the range of scientific knowledge, policy guidelines and legal controls in the context of local conditions and specify the nature and degree of control measures required. When the local authority is of the type of the Alkali Inspectorate or Regional Water Authority, with adequate scientific and technical expertise for the job, this task is difficult enough. When it is the county or district council, with a range of responsibilities over and above pollution control, difficulties arise due to lack of manpower and expertise and control may be less than adequate. In addition, Local Authorities may find themselves lacking sufficient guidance from central government for the implementation of their responsibilities for pollution control.

A recent case illustrates this point. Cheshire County Council tried to incorporate into its Structure Plan,[2] the WHO target levels for smoke and SO_2 as the ones against which the Authority should judge applications for permission to develop industries which might cause this kind of pollution in their area. The Council felt that these guidelines would provide a more useful evaluative tool than the present government guidelines, which merely state that pollution considerations are important in reaching planning decisions. DOE, however, in response to this initiative announced that:

> "the Secretary of State accepts the panel's[3] recommendation that in deciding on the acceptability of a given level of air pollution the planning authority should have regard to available scientific. evidence. It would, however, be inappropriate for him to approve policies based on standards which have not been accepted nationally and which are likely to give rise to difficulties in implementation".

The local authority were thus left with the existing guidelines, despite the clear demonstration of their preference for more detailed and specific criteria upon which to base control decisions and enforcement.

The multiplicity of groups involved in pollution control in each "system" and at different levels of government has implications for both the collection, provision and exchange of information. National monitoring schemes may provide quite adequate information for the compilation of a national picture to serve national assessment and policy formulation, but provide data of insufficient detail for local control and enforcement purposes. This is the case, for example, with the National Survey of Smoke and Sulphur Dioxide.

Comprehensiveness, accuracy, and cost effectiveness are important features of monitoring programmes. Because of the enormous costs involved as well as the manpower requirements (both in terms of numbers and expertise), information required for assessments may be inadequately collected, or essential monitoring surveys deferred due to unfavourable economic conditions. The accuracy of different measurement and analytical methods, compatability of

---

[2] Structure Plans are made by County Councils setting out, in broad terms, the nature of development to be permitted in the county over the next ten years. They must be ratified by the Secretary of State for the Environment.

[3] The Panel of Enquiry set up by DOE to examine, in public, this issue and recommend a suitable solution to the difference of opinion between the County Council and DOE.

different data records and standardization of final results to permit inter-comparison, are also important considerations for the cost-effectiveness of monitoring surveys. The numerous groups involved in pollution control in different capacities may often result in situations where different departments employ different methods when monitoring similar or related pollution problems, and render comparison of data, and its multiple use, difficult.

Confidentiality is a further obstacle to information provision and use, and affects human health and industrial data particularly. In many cases, in the UK for example, industry monitors itself, with the control authority acting in the role of "back-up" monitoring, for enforcement checks. Information collected by industry is almost always confidential between the industry and authority or department involved. Health data are never released in anything but aggregated form to guard personal privacy. The problem of confidentiality occurs in many countries.

Conclusions

While this paper has examined the case of lead pollution control, some general conclusions can be drawn with regard to risk assessments of pollution. One important point which emerges from this study is that risk assessment methodologies developed in isolation from the needs and realities of control run the risk of being unusable in such contexts. This is inevitable because of the institutional constraints on the provision of the necessary expertise and information required to carry out assessments; as well as the charcteristics of strategies for pollution control which pose different requirements for assessment results.

The need for improved monitoring information for risk assessments has already been recognized in the U.K. There are, at central government level, inter-departmental groups examining the feasibility of harmonizing the monitoring activities of the numerous groups involved in pollution control. Their task has not proved easy to date. Interest in the training of pollution control officers is also evident. But economic considerations are always strongly influential on the quality and pace of change and improvement.

It is also clear from this study of lead pollution control that, in risk assessments of energy-related problems, pollutants cannot be considered in isolation. The nuclear case apart, almost all the pollutants released through energy exploitation are also released to the environment from other, non-energy, sources. Some not as widely as lead perhaps, but they are certainly released from non-energy sources in sufficiently significant quantities to make the consideration of the total range of sources and transfer processes an essential requirement in any assessment.

A pragmatic, case-by-case approach to pollution control will, for the most part, have different assessment requirements from one based on fixed standards and uniform emission control. BPM is often regarded as uniquely British, this is not so. The same, or similar, criteria operate in other countries, including Australia, Canada, Federal Republic of Germany and the U.S.A. For example, the U.S. Federal Water Pollution Control Act 1972 (Amendment) states that:

> "Factors relating to the assessment of *best practicable control technology* currently available ... shall include consideration of the total cost of the application of technology in relation to effluent reduction benefits . . ."

as well as the age of equipment, process employed, etc.

Furthermore, the characteristics of the British control arrangements which can influence risk assessment and vice versa are also evident in the control arrangements of many other countries. A sectoral approach, with numerous departments and groups involved in control is almost inevitable everywhere. Where there are departures from this general pattern, they are normally the result of fairly recent changes, such as the establishment of the U.S. Environmental Protection Agency (EPA), in 1970. But even this type of development only partially changes the situation. There are still a multiplicity of groups involved in environmental protection in the U.S.A. in addition to the EPA.

Whatever the details of approach, it is undeniable that those administering controls are in continual need of reliable criteria and guidelines from assessment with which they can effectively enforce legal control provisions. Often, upon examination, it is found that there are sufficient legal powers available to deal with the majority of pollution problems, but because of inadequate assessments, these remain under-exploited. The early warning of potential pollution hazards is also a major concern of governments. But, scientific uncertainties apart, how are such assessments to be carried out in the framework of existing control structures, without imposing impossible organizational and financial burdens on government authorities? The methods for assessing the chronic effects of pollutants are also in need of improvement so as to provide workable techniques for control. These are the type of questions which must be addressed by scientists concerned with risk assessment, who wish to see positive action being taken in response to the results of their assessment by government.

There is a need for closer co-operation and improved methods of working, between scientists and administrators. Language difficulties play their part certainly, but attitudes on both sides could also bear examination. Scientists, if they do not firmly grasp the opportunity to contribute positively to pollution control in the broadest sense, may also lose the respect and confidence of the public, and thereby further miss an opportunity to promote improved pollution control and environmental protection by informing and influencing public opinion and thereby government policy and action.

Acknowledgements

The author acknowledges financial and technical support from the U.K. Department of the Environment. However, the views expressed in this paper are the author's own and do not necessarily represent the views of the Department of the Environment.

References

Commission of the European Communities (1976). "State of Environment Report". Commission of the European Communities, Brussels.

Goodman, G.T. et al. (1976). The use of biological materials as environmental pollution gauges. In "Proceedings of the International Colloquium on the Evaluation of Toxicological Data for the Protection of Public Health". Pergamon Press, Oxford.

O'Brien, B.J. (1979). "The Exposure Commitment Method with Application to Exposure of Man to Lead Pollution". MARC Report No. 13. Monitoring and Assessment Research Centre, London (in press).

O'Brien, B.J. and Coleman, D. (1977). "Lead in the Global Environment", Monitoring and Assessment Research Centre, Report No. 12, London.

U.K. Department of the Environment (1974). "Pollution Paper No. 2: Lead in the Environment and its significance to Man". Her Majesty's Stationery Office, London.

U.K. Department of the Environment (1977a). "Pollution Paper No. 9: Pollution Control in Great Britain – How it Works". Her Majesty's Stationery Office, London.

U.K. Department of the Environment (1977b). "Pollution Paper No. 11: Environmental Standards, A Description of United Kingdom Practice". Her Majesty's Stationery Office, London.

U.S. Environmental Protection Agency (1971). "Environmental Lead and Public Health" (publication no. AP-90) Air Pollution Control Office, Research Triangle Park, North Carolina.

World Health Organization (1977). "Environmental Health Criteria 3: Lead". World Health Organization, Geneva.

POLITICALLY ACCEPTABLE RISKS FROM ENERGY TECHNOLOGIES: SOME CONCEPTS AND HYPOTHESES

Evert Vedung

Department of Government,
University of Uppsala, Sweden

Three Types of Risk Studies on Energy

In the societal process of balancing risks and benefits from energy sources and technologies, three types of risk assessment studies merit special attention.

Type 1

It is generally taken for granted that full and reliable information concerning every conceivable consequence of new and unadopted energy technologies is quite impossible to acquire. However, there is a distinct trend towards making assessments of consequences much more formal and technical than before, as evidenced by governmental programmes and agencies for technology-assessment. Thus, new technologies and products are increasingly subjected to scrutiny through several types of formal testing procedures. Scientists and technicians from a wide variety of disciplines are called upon to make expert assessments of probable consequences and provide information on different aspects of new technologies. Technical reports, fault-tree analyses and cost-benefit analyses are all used to provide formal and rigorous assessments of the probability, scope, intensity, and actual incidence of particular risks from energy sources and technologies. This type of risk assessment study is guided by the following questions:

1) What is the problem to be considered?
2) What alternatives are available for solving the problem?
3) What outcomes are associated with each alternative? What are the probabilities of these outcomes?

However formal and rigorous these assessments may be, they can only provide information about what the risks *are*, and not what the *socially acceptable* levels of risk should be. They thus constitute only one part of the overall societal calculus with regard to the acceptability of new, potentially hazardous energy technologies.

Type 2

This is where the second type of energy-assessment study, the *normative* one, comes to the fore. Compared with the first, two more operations must be taken into account. First, the perceived consequences must be evaluated by means of some system of preferences or values. Such a preference system or utility function must be thought out and adopted. The next step would be to ask the question, which energy technology ought to be adopted given the preference system and the information about consequences of each alternative. Second, a decision rule is also needed in order to reach a decision concerning what the level of risk should be. In the literature on game theory, decision theory and political evaluation, several different decision rules are presented and discussed. There is no general agreement in the literature as to which rule ought to be applied. Some advocate the rule that the alternative which promises to yield the maximum possible payoff should be chosen (the maximax principle). Others prefer a more conservative criterion and favour the rule that specifies choosing the alternative which maximizes the minimal gain (maximin principle). Still a third decision rule is the principle of insufficient reason. It says that all states of nature ought to be given the same probability if you cannot be sure what the actual probabilities are. The alternative which seems to yield the highest expected value should then be chosen.

To sum up, the normative type of risk assessment study is guided by a fourth and a fifth question, in addition to the three posed above in connexion with the first type of assessment, *viz.*:

4) What are the preferences to be applied?

5) What alternative should be chosen, given consequences, preferences and a decision rule? In other words, how should calculi three and four be integrated and intercalibrated with a decision rule in order to reach a final decision with the most favourable overall results.

Type 3

The third approach to risk assessment is to study how acceptable levels of risk are *actually determined* in various sectors of society. In contrast to the two preceding ones, this is a study of decisions *ex post*, i.e. decisions which have already been made. What are the factors that explain why one level of risk is deemed more acceptable than another?

This paper presents some comments on how empirical risk studies of this third type can be carried out. The focus is on the political aspect of determining acceptable levels of risk from energy technologies. The rationale for this follows.

The Political Aspect of Risk Assessment

In the past, many decisions about new technologies in the energy field were taken in the market-place. The early decision to exploit and use oil as a major fuel was made at a very decentralized level by producers and consumers of the product. No grand decision was ever made by nationally elected politicians to adopt oil as a major source of energy.

The innovation of hydroelectric power is an example of a process with more political features in it. Waterpower was introduced on the market by private entrepreneurs. After some decades, government decided to regulate further exploitation by law. It also decided to build, own and run water-power stations and a distribution system for electric power. But private companies have remained in the field, and we now have a mixture of government and private enterprise.

In modern times, the adoption of new technologies has gained an even stronger political aspect, at least in Sweden. This depends on the fact that new technologies, such as nuclear power, seem to offer very high benefits and very high risks as well. The use of some new technologies seems to take on increasingly far-reaching societal ramifications. It leads to incidences of risks and benefits on scales which more and more involve overarching and perennial political issues such as equality, freedom and justice. Decisions about adoption of new technologies which might cause ecological catastrophes, it is argued, must not be made in the market-place. They must be taken in the political domain. Thus, the immense ramifications of adopting a potentially risky technology make the political perspectives and belief systems of ordinary voters and nationally elected politicians more and more crucial for the study of how acceptable levels of risk are actually determined.

This kind of study can be conducted on the mass-level of the political system; one can ask how ordinary voters make up their mind on questions of risk. But it can be conducted on the level of political elites as well. This means that the decision processes of top politicians, administrators, business-leaders, pressure-groups, researchers and media people can be investigated. In the sequel, I will concentrate on the latter level, or more particularly, on the level of political parties.

The Rational Choice Model

The study might be guided by different models of description and explanation. I will draw attention only to one which I will refer to as "rational choice model". It consists of the following questions addressed to the material at hand:

1) What was the problem considered by the "actors"?

2) What alternatives for a solution were available?

3) What substantive outcomes were associated with each of the available alternatives? What were the probabilities of these?

4) What strategic outcomes were associated with each of the available alternatives? What were the probabilities of these strategic outcomes?

5) What preferences were applied to the substantive outcomes?

6) What preferences were applied to the strategic outcomes?

7) How were 3, 4, 5 and 6 combined with a decision rule in order to reach a final choice?

Questions four and six need some comments. We can see immediately that they are added to those five which were said to be crucial in a risk study of type 2. They deal with strategic or tactical considerations. The fourth question includes considerations about the probable behaviour of other political "actors" — such as other parties, individual politicians, pressure groups or members of the electorate — if a certain alternative is chosen. If a political party adopts one line of action instead of another, this might be thought to diminish electorate support and weaken internal party cohesion, but increase the possibilities of joining a future coalition government. In other words, the position of the party is considered to be influenced by the moves of other players in the game. The sixth question is about how to evaluate the various perceived strategic consequences and outcomes. These two strategic calculi are of utmost importance when it comes to explaining the choice of alternative in a game with more than one player. And, in politics, this is always the case. Not even dictators are in such a strong position that they can entirely neglect the strategic implications of their decisions.

The Problem and the Alternatives

One celebrated issue in Swedish energy policy is the question — which primary energy source or, rather, which combination of primary energy sources is to be concentrated upon? In principle, government could choose between extracting energy from:

1.	waterpower	9.	ocean heat
2.	wood	10.	biomass
3.	waste products	11.	crude oil
4.	waste heat from paper mills	12.	natural gas
5.	wind	13.	coal
6.	solar energy	14.	peat
7.	ocean tides	15.	nuclear fuel
8.	ocean waves	16.	geothermal power

Viewed from a risk-assessment angle, the *problem* might be described as one of choosing the combination of energy sources which yields the most acceptable risks. Three major forms of response seem conceivable to decision-makers faced with the responsibility of deciding politically-acceptable levels of risk from energy production:

1) measures to accept consequences by bearing or sharing the consequences;

2) measures to reduce or change probabilities and/or consequences of events; and

3) measures to avoid the risks, totally or for a period of time.

These responses can be considered as the *options* available. They are summarized in Figure 1.

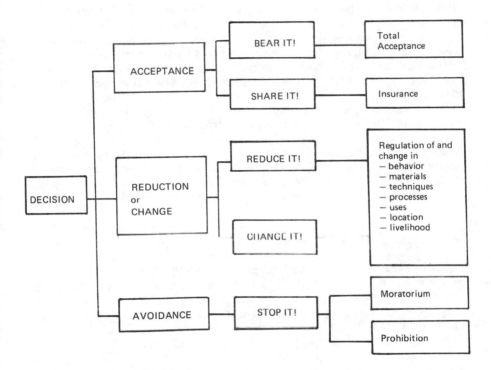

Fig. 1 A classification of conceivable measures for coping with risks.

Explanatory Hypotheses

The next step is to explain why different decision-makers tolerate different risk levels within one and the same risk sector or why all decision-makers tend to tolerate greater risks within one sector than in another. It seems that the following explanatory factors must be taken into account:

1) Factors concerning the nature of the risk source.
2) Factors concerning the reactions of the risk-receptor.

3) Factors concerning the relation between the nature of the risk source and the receptor.

4) Preferences and ideologies of politicians.

The hypotheses presented below are organized according to this scheme:

Hypotheses Concerning the Nature of the Risk Source

New Versus Old Energy Technology It is reasonable to assume that risks of a new energy technology will not be accepted in the same way as risks belonging to an established technology or activity. Regarding established activities, one has already committed oneself to certain sacrifices and is already receiving benefits which were given at the outset. This inertia, or "tiedness," together with the many obvious benefits, makes it easier to accept the risk of an already established technology than, e.g. the same level of risk associated with a new activity which is replacing the older one.

Known Versus Unknown Probabilities It seems that people more easily accept risks with known probabilities than risks with unknown probabilities. Known probabilities are based on a statistical "mapping"; unknown risks are based at best on theoretical "mapping".

Hypotheses Concerning the Reactions of the Risk-Receptor

Voluntary Versus Involuntary Exposure People are inclined to take larger risks in voluntary activities than in involuntary activities, i.e. when risks are imposed upon them. For example, we know from existing data that people engaging in sport or other leisure activities seem prepared to take about 1000 times larger risks than those imposed by public transport for instance. This means that voluntary risks, which people personally decide to take after some kind of intuitive calculus of costs versus benefits are more readily accepted than are involuntary risks which are entirely controlled by some external authority.

Distributed Versus Non-Distributed Risks People more readily accept risks which are diffuse and distributed over a larger number of people than such risks which are clearly localized to a small group.

Conventional Versus Unconventional Damages People would rather accept risks which give rise to immediate and conventional damage than those which give delayed effects, such as foetal or genetic damage.

Hypotheses Regarding the Relation Between Risk-Source and Risk-Receptor

Small-Effect Accidents That Happen Frequently Versus Large-Effect Accidents That Happen Seldom More numerous, but smaller risks are more easily accepted than fewer risks with larger effects (e.g. road- versus air-travel). This valuation probably relates to the fact that for any given activity, risks with small effects occur more frequently than risks with large effects. An important corollary of this is that society generally seems to take the view that *there is a tolerable limit to consequences, regardless of how low the probability is for consequences of unlimited size.* To consciously or intuitively set this limit is an act of value-judgement which is built on the benefit of the activity. For example, "It must never happen that a whole society is wiped out if a water-dam breaks." Theoretical arguments about probability are not relevant here.

Risks With Very Low Probabilities If the probability of an incident with dangerous consequences is extremely small, the risk is usually disregarded, no matter how large the possible effect. As the probability of a risk gets smaller and smaller, the group or society sets a limit, beyond which the risk is regarded as given by "fate" rather than as a result of human activities. The risk thus becomes negligible. This level of probability is the same as the one concerning large natural catastrophes, e.g. you do not build a roof over a sports-stadium to withstand meteorites. The judgement is that such a risk can be neglected.

Cause and Effect If an accident has happened, we always ask if it could have been avoided. Could we, in advance, have broken the cause-effect chain by restructuring it, by better control, or by more education? Or should the activity even be forbidden? In considering a new technology, the same type of questions apply. Natural catastrophes or climatic conditions are examples of causes which cannot, *per se*, be eliminated by human management. For incidents with relatively trivial effects, it is enough that all reasonable measures to break the cause-effect chain are taken. But for cause-effect chains with more serious consequences, i.e. where the target of the effect can be recognized in advance and where the effect is unambiguously explained by the activity, people demand that the cause-effect chain be broken, either through forceful measures, or by forbidding the activity altogether. Here we come close to the idea of an absolute ban on consequences, regardless of their very low probability, as a result of societal value-judgements.

Widespread Technology Affecting Many People or Minor Technology Affecting Only a Few The more widespread technology becomes, i.e. the more people affected by the technology, the less individual risk is taken. The risk is transformed from being a matter of concern for an individual to being a matter of concern for a sector of society or the population as a whole. This evaluation can be illustrated by the development of, e.g. private cars, air traffic or electrification. The reduction of risks depends partly on the fact that the technology is developed according to the users' demand for improved safety, durability etc., and partly on how widespread the technology is. When the activity becomes widespread, it is obviously in the best interests of society to protect the life and limb of its members, e.g. through legal measures. The more rapid the spread of the technology, the greater the demand for improved rules to handle it. This requires planners to consider the future, as well as the present, requirements of the technology. The lower willingness to take risks when a technology becomes widespread is also connected with a degree of control which the individual can exercise over it.

Changed Benefit, Changed Risk Up to a certain limit, individuals will take on risks that grow faster than the growth of the corresponding benefit. This connexion is true, for instance when a sick person evaluates a new but uncertain method to cure a previously intractable sickness. It is doubtful, however, whether this connexion can be directly applied to whole groups of people.

Hypotheses Concerning Preferences of Politicians

Let us imagine a situation in which three politicans, A, B and C, are faced with the task of making a decision concerning the politically-acceptable level of risk for energy technology T. The main consequence of T is perceived by A as being positively linked to a dimension of high centrality in his belief system. Politician B, on the other hand, perceives the same consequence as being contrary to his central values and beliefs. For politican C, the consequence of T seems contrary to one, but consistent with another of his most central values and beliefs. Furthermore, politician A perceives the incidence of risk as falling outside of his electorate, while B views the incidence of risk as falling almost completely on his supporters. Politician C, finally, draws his support from a wide spectrum of groups in society. Consequently, he perceives some of his support groups to be exposed to the risk, while others remain untouched or even receive benefits from T.

In this situation, it is tempting to hypothesize that politician A will cast his vote for the acceptance of T. His tendency towards T will be strengthened by his perception of the strategic consequences of adopting T: thanks to the incidence of risks and benefits, his support groups will thrive from T, and thus continue to support A. Politican B, on the other hand, is likely to request a moratorium or a prohibition of T. This propensity of B – which is already high as a result of B's preference calculus, will increase as a result of B's perceptions of the incidence of risk as coinciding with his most important support group. Again, C is in a precarious dilemma. We may assume that he is likely to opt for an alternative involving reductions and changes in the risks associated with T, but he might as well ask for a moratorium in order to have a closer look at available information or to wait for new information crucial to his dilemma.

We thus assume that differences in the final choice of politically-acceptable risk levels are intimately and systematically connected with the perspectives and belief systems of political decision-makers.

Acknowledgement

The first four sections of this paper were originally written in close co-operation with Lennart J. Lundgvist of the Department of Government, Uppsala University. I wish to thank him for his inspiring contribution and absolve him from any responsibility for this version of the research note.

PART 5
THEORY AND APPLICATION

INTRODUCTION

There are two fairly clear-cut stages in comparative risk-assessment. The first is the technical analysis of the range and severity of risks associated with each of the activities or technologies from which a choice has eventually to be made. In the case of the risks associated with the various energy-options, this involves the collation and analysis of scientific and technical reports bearing on the risk-profiles of each of the options concerned. These contain actuarial, epidemiological, ecological and other statistical data, as well as occupational-health, toxicological and other experimental results. Fault-tree analysis and modelling are often made use of here.

The second stage reviews the results obtained from the first stage against the existing socio-economic and cultural value-systems of the society concerned, in order to obtain the "best-bargain" between minimal risk and societal needs and values. This stage includes cost-benefit studies, environmental impact analysis, modelling, decision-analysis and policy-analysis. The first stage approximates to "Risk-Identification" and "Risk-Estimation", and the second stage, to "Risk-Evaluation", as explained by O'Riordan in Part 1. The results of these two stages are then taken over by governmental and other executive bodies for implementation. This requires regulation, control and legislative-enforcement action, often with some form of monitoring activity to evaluate the efficacy of the control measures being implemented. This is O'Riordan's "Risk Control" phase.

Many of the scientific and technical activities associated with the first stage have already been discussed in Parts 1-3, leaving some of the second stage activities to be dealt with here. Cost-benefit analysis and modelling have been selected for treatment, along with some aspects of implementation, notably institutional problems connected with government management of risk and the political aspects of decision-making over risk.

Cost-benefit analysis has often been welcomed as a valuable tool been asked and the analysis made visible.

Part 4 may have left the reader feeling that in their present state of development, formal risk-assessment methods are able to play little or no part in servicing the practical, everyday risk-management of the future. Is this pessimistic view justified? There is no doubt that the systematic analysis *post hoc* of the processes of evaluation and decision-making as done by Fischer and by Pearce for north-sea oil and THORP, – issues that have already been decided – can help point out what should have happened in a way which is altogether far more illuminating than simply being wise after the event. In fact, every contribution in the book makes valuable retrospective analyses of one kind or another. The real challenge for the study of risk-assessment is how to marshal this knowledge to create a process-orientated approach with predictive capability for future risk-situations, thus enabling the most appropriate policy-options to be selected for implementation. Most fields of study have to go through this phase in their evolution to becoming well-ordered disciplines and the study of risk is no exception. The greatest stimulus and most rapid progress will come from practice. Risk-assessment practitioners should be given every opportunity to become involved in decision-making processes as they occur in order to develop a more applicable field-theory for future use.

WHAT IS AN ACCEPTABLE RISK AND HOW CAN IT BE DETERMINED?

W.D. Rowe

Institute for Risk Analysis, The American University, Washington, D.C., U.S.A.

Individuals make decisions involving voluntary exposure to risks on a continuing basis. In many cases these decisions are *de facto* since the risks are discounted or ignored. In the first case, an individual may perceive that a statistical risk will not happen to him (airplane crash), that he has some control in avoiding unwanted events (driving an automobile), or that the benefits to his quality of life far outweigh any consideration of risk (addicted cigarette smoker). In the latter case, anxiety involved in thinking about risk is minimized.

While there are lessons to be learned from examining individual voluntary risk decision behaviour, primary interest in risk assessment activities is focused upon societal decisions involving the imposition of involuntary risks on parts of society from technological, institutional, and business activities. Thus, the determination of acceptable levels of risk associated with an activity is a social process involving the balancing of costs, risks, and benefits whose distribution is often inequitable. Recognition of the difficulties in quantifying social variables and the impossibility of unanimous agreement on any social issue is a prerequisite for understanding what makes a level of risk acceptable.

Unquestionably, some risks are acceptable. Some conditions that support this contention are evident:

1) A risk is perceived to be so small that it can be ignored – *threshold condition.*

2) A risk is uncontrollable or unavoidable without major disruption in lifestyle – *status quo condition.*

3) A credible organization with responsibility for health and safety has, through due process, established an acceptable risk level – *regulatory condition.*

4) A historic level of risk continues to be an acceptable one – *de facto condition.*

5) A risk is deemed worth the benefits by a risktaker – *voluntary balance condition.*

There have been many definitions of acceptable risk, unacceptable risk, unreasonable risk, etc. Most of them involve the methods of establishing references for comparing risks such as revealed, implied, and expressed preferences (Kates, 1977), or evaluation of a spectrum of possibilities for different situations, (Lowrance, 1976; Council for Science and Society, 1977).

To circumvent all the arguments involved in establishing a precise definition, a working or operational definition may be useful: *A risk is acceptable when those affected are generally no longer (or not) apprehensive about it.*

As a general statement, it does not mean *all* those affected. Nor does it matter by what process the level of acceptability is achieved. It does however, include the "regulators" and "experts," since they must be satisfied that the risks are low enough. It relates to the propensity for anxiety aversion as well as the propensity for risk aversion of society. It addresses voluntary and involuntary risks. The broadness of the definition reflects that there is no single, universal method of arriving at an acceptable level of risk.

Unacceptable risk must also be defined. An approach is shown in Figure 1. As risks increase in magnitude from zero they are initially

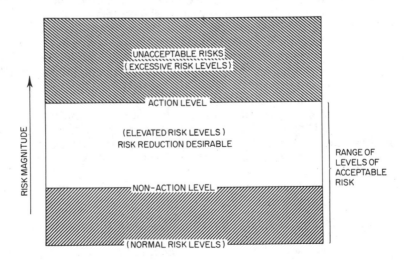

Fig. 1 Levels of interest for setting acceptable risk levels.

at a "normal" level below some "non-action" level. These risks are considered so small as not to be worth spending administrative time and resources on. Risks become elevated as their magnitude increases until an "action" level is reached. Risks above this level are "excessive" and unacceptable, and action to reduce them is *required.* Elevated risk levels between "non-action" and "action" levels must be evaluated in each case for setting an acceptable risk level.   The range for acceptability is shown on the right hand, vertical axis. Action to regulate a risk may start at any place above the "action" level;   however, the resultant level may be below the "non-action" level. That is, an initial level of risk may or may not invoke regulatory action, but once regulatory attention is invoked, other factors must be considered to set an acceptable level of risk.

Before addressing the problem of how acceptable risk levels may be determined under such a definition, the nature of the underlying process of how society makes choices, or avoids making them, must be better understood. For this purpose, a "Theory of Social Non-Choice" is offered to provoke discussion.

A Theory of Social Non-Choice

Present economic and social decision theory is based upon the precept that in a society many decisions affect large numbers of people, and that the preferences of the individuals concerned can be ascertained.   Various methods for obtaining social choices from individual values have been advanced, particularly the work of Arrow (1963), Coombs (1954), and Farris and Sage (1975). These ideas have evolved into a Theory of Social Choice in which the conflicting interests of collections of individuals must be resolved by conflict or compromise. The theories attempt to take individual preferences and develop these into social welfare functions by determining an optimum group preference structure from the alternatives. In these cases, each individual is expected to be willing to express his preferences for fulfilling his wants and desires.

The basis for these approaches is the assumption that people will act to obtain (or at least express) their preferences in proportion to the magnitude of their perceived utility measure for each preference. This linear relationship is illustrated in Figure 2 over a limited range from zero to some value of utility. This range covers the realizable range of utility in that the choices are realistically achievable. As utilities increase outside of this range, individuals have little experience to go on in making such judgements and the general relationship becomes uncertain and probably non-linear in some direction. This is illustrated by the shaded area in the Figure. It may well be that such expressions of preference do take place for satisfying needs.   However, if the reverse of preferences, that is undesirable consequences (threats), are used as a measure, the process breaks down.   The avoidance of unwanted consequences is equivalent to the process of averting risks in society. When stating preferences for avoiding or reducing risks, a new factor, anxiety about risks, is introduced.   People want to think about desirable things, even if unattainable, but they avoid anxiety caused by thinking about undesirable things to the extent possible.   Thus

W.D. Rowe

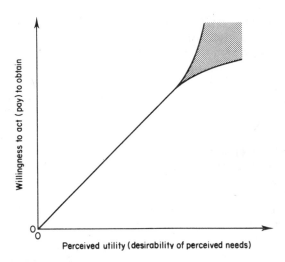

Fig. 2 Relationship between perceived utility and willingness to act for objectives of worth.

behaviour to avoid risk does not follow the same pattern as obtainment of wants and desires. This is illustrated in Figure 3 where each scale is a negative, mirror image of those in Figure 2. An anxiety

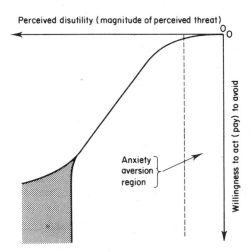

Fig. 3 Relationship between perceived disutility and willingness to act for undesirable objectives.

aversion region is shown as derived from a non-linear process. It is based upon the percept that people generally avoid anxiety by suppressing concern about threats unconsciously (or even consciously). Such anxiety avoidance has often been identified and investigated by psychologists.

It is postulated here that society is anxiety averse, i.e. individuals *and* groups attempt to minimize anxiety. Generally, an individual is faced with many real or perceived threats at any given time. Unless paranoid, he will usually ignore most threats on a conscious basis, a reflex action to prevent anxiety overload. Only when a threat or stress is large enough to exceed some threshold level is conscious anxiety aroused. Depending on the nature and magnitude of the threat, he will suffer in quiet anguish or choose one of Bion's (1952), four responses: fight, flight, dependency or pairing. Thus, depending upon the nature and magnitude of a stress, three kinds of action occur which are not necessarily mutually exclusive:

1) Ignore threats on a conscious or unconscious basis to avoid anxiety.

2) Consciously suffer the stress and remain in a state of anxiety.

3) Act to reduce the stress and anxiety.

The asymmetrical relationship between utility and willingness to act for positive and negative utility is shown in Figure 4, which

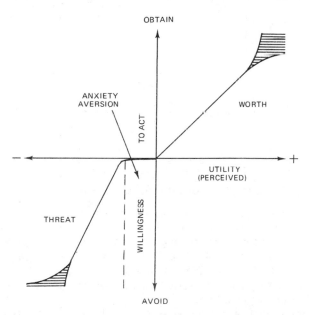

Fig. 4 Non-symmetry between actions to obtain or avoid desirable and undesirable objectives.

combines the two previous figures into a single illustration. The anxiety aversion of society, so illustrated, provides a basis for insight into the reasons why large numbers of people in society are indifferent to many of the threats seemingly imposed on them. It is evident from voter response and public participation in forums that the general population is generally indifferent to societal problems whose direct effect on individuals is minimal. This lack of interest on parts of the population has often been called a major defect in our democratic society, i.e. individuals have not exercised the responsibilities that accompany the rights of a democratic society. Anxiety aversion may be one cause.

In any case, on a given threatening issue, members of society divide into five more or less easily identifiable categories as shown in Figure 5. These are:

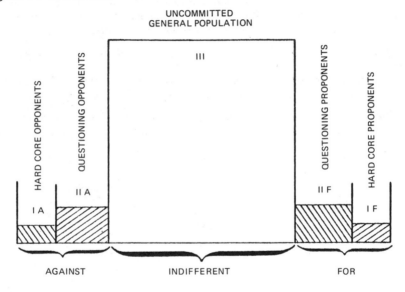

(AREA IN EACH CATEGORY IS AN APPROXIMATE ESTIMATE OF THE RELATIVE POPULATION SIZES)

Fig. 5 Categories of adherents to aspects of issues.

1-A (Against): Hard core *opponents* to an issue whose opposition is based on a belief, rational or irrational, that the issue is wrong.

1-F (For): Hard core *proponents* of an issue whose support is based upon a belief, rational or irrational, that the issue is right.

2-A (Against): Questioning *opponents* who have doubts about the issue and want to postpone action until the doubts are cleared up by rational answers to problems.

2-F (For): Questioning *proponents* who have doubts about the issue and are amenable to continued action while rational answers are pursued to clear up outstanding doubts.

3: Uncommitted general population who are indifferent to the issue unless their anxiety threshold is exceeded.

A first approximation of the relative numbers on an unexploited issue is also shown in the Figure.

Group I positions are taken on an issue, based upon belief, for or against.   Once such a belief is established it is very difficult to change that belief.   These people are already committed.   Group 2 positions are based upon doubts, but doubts that can be assuaged by serious action to resolve the conditions from which the doubts arise. Before an issue has escalated, the general population and even the media do not give too much credence to Group I positions.   They are usually considered extreme and often fanatical.   However, questioning opponents or proponents, especially those with recognized scientific or expert credentials, are given attention.   If their doubts are not assuaged, the issue can become polarized and will escalate.   This is illustrated in Figure 6 where continued lack of resolution has caused

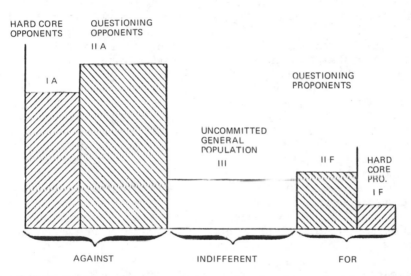

Fig. 6 Polarization of adherents toward major issues lacking credible resolution.

concern to the uncommitted, leading them to join the ranks of the opponents, hard core and questioning.   Nuclear energy is an example of an issue which has so escalated.

The Theory of Social Non-Choice is based upon the thesis that press coverage and credible attention is given to Group 2-A, (credible questioning opponents), out of all proportion to the numbers in the population.   Group 2-A members are credible, will be listened to and their ranks will swell if believable answers to unresolved issues are not forthcoming.   Thus, the Theory of Social Non-Choice proposes that it is this group which makes choices for society, not the general population as a whole.   Acceptable solutions to issues must be addressed to this group, and if addressed fairly, will satisfy questioning proponents as well. If these groups accept a solution, then the fears of Group 3 members will be assuaged and they will continue to be indifferent.

One conclusion that may be drawn from this hypothesis is that early attention to addressing issues brought up by Group 2-A activists may prevent an issue from escalating. This means that when problems can be resolved early in the existence of an issue, most of the public will not be concerned nor do they want to be. The object is to provide a credible means to address such issues. A government regulatory agency with a responsibility to protect the public and a credible process for doing so is one means of accomplishing this end. Establishing acceptable levels of risks for technological systems is one such approach.

Is such an approach paternalistic? If one believes it possible to make better decisions than the population, the answer is yes. But if one is acting in place of an uncommitted population who refuses to participate, there is no other way. It is not that a regulator is wiser than the population, it is just that he must act for people who refuse to participate and who have abrogated their responsibilities. Thus, such an approach need not be paternalistic.

Is such an approach manipulative? If one acts to restrict participation, the answer is yes. But, if one acts fairly to address outstanding questions and their resolution, dealing with them openly, and with active public participation always in mind, it is the only way to assure adequate review and discussion; and the process need not be manipulative.

There is always the dilemma that early attention to identifiable risks may give credibility to concern about them prematurely. The separation of legitimate concern from posturing to support unproven beliefs is difficult, but not to recognize such differences early in the process leads to loss of credibility.

Approaches to Determining Acceptable Levels of Risk

Resolving issues by setting levels of risk by credible, demonstrable, and visible processes that are judged to be low enough by Group 2 and thereby Group 3 standards can reduce apprehension. Thus, any such levels are deemed as "acceptable" by the operational definition given previously. The question now to be addressed is: How can they be determined?

A regulatory agency setting standards has a variety of processes available for approaching a particular decision on risk acceptability. These involve comparison with risk references obtained by:

1) Examination of individual and population risks by revealed, imputed, or expressed preferences.

2) Examination of the costs of risk reduction for individuals and populations.

3) Balancing on a gross basis cost, risks, and benefits for equitable

risks.

4) Case-by-case examination of each decision using any combination of the above.

Examination of Individual and Population Risks

Exposure to a threat involves the probability that an undesirable consequence will result to an individual or population. The relationships between exposure and consequence take many forms as shown in Figure 7. Curves 1 and 2 show threshold and breakpoint relationships,

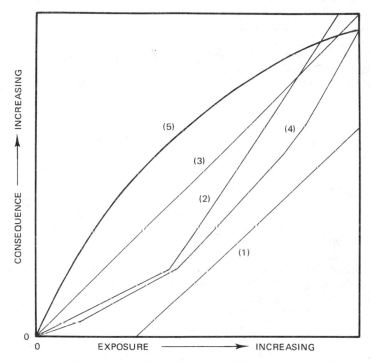

Fig. 7 Relationships between exposure and consequences (from Rowe, 1977a, with permission).

respectively. These abrupt points of change provide direct criteria for establishing acceptable levels of risk below the breakpoints, perhaps with "an ample margin of safety." The remaining three curves are continuous down to zero, with curve 3 the linear, non-threshold relationship. These latter curves have no internal (i.e. from the shape of the curve alone) criteria for establishing risk acceptability. External criteria are required.

Comparisons using similar risks in society as risk references is one approach in establishing levels of acceptable risk. Clearly it is important that risks of similar nature be compared. Thus, voluntary risks of accidents from driving automobiles cannot be compared with cancer risks involuntarily imposed on a portion of the population (Rowe, 1977a). The real concern is with risks imposed inequitably, including intergenerational inequities.

Three approaches to obtaining such references are (Kates, 1977):

Revealed Preferences This method is based on the assumption that, by trial and error, society has arrived at a nearly optimal balance between the risks and benefits associated with any activity. One may, therefore, use statistical cost, risk, and benefit data to reveal patterns of acceptable risk-benefit trade-offs. Acceptable risk for a new technology is assumed to be the level of safety associated with ongoing activities having similar benefit to society.

Expressed Preferences The most straightforward method for determining what people find acceptable is to ask them to express their preferences directly. The appeal of the expressed preference method is obvious. It elicits current preferences, thus being responsive to changing values. It also allows for widespread citizen involvement in decision-making and thus should be politically acceptable. It has, however, some possible drawbacks which seem to have greatly restricted its use. For example, people may not really know what they want; their attitudes and behaviours may be inconsistent; their values may change so rapidly as to make systematic planning impossible; they may not understand how their preferences will translate into policy; they may want things that are unobtainable in reality; and different ways of phrasing the same question may elicit different preferences.

Implied Preferences The implied preference method may be seen as a compromise between the revealed and expressed methods. It looks to the legal arrangements in a society as a reflection of both what people want and what current economic arrangements allow them to have. Its proponents, like those of the democratic process, make no claims to perfection; rather they see it as a best possible way of muddling through the task of bringing risk management in line with people's desires. The problems here are familiar to any participant in a democracy: our legal arrangements include not only laws adopted by our elected representatives, but also interpretations, precedents and improvisations by judges, juries, regulators, and others; laws are sometimes written poorly and inconsistently; often they are extended to cover situations undreamed of when they were written; and their precise formulation often reflects fleeting political coalitions or public concerns.

The difficulty of setting acceptable levels of risk by considering risk alone is not only confounded by inequitable risk distribution, but by the size of the populations involved. Risk to an identified risktaker is different from that to a statistical member of the population (see Rowe, 1977a). Large risks to a few people and small risks to large numbers of people are not directly reconcilable, and concern for both aspects requires widely different approaches in their consideration.

Cost Effectiveness of Risk Reduction

When risk criteria alone are inadequate to establish acceptable risk levels, economics may be brought into consideration. This results in the cost-effectiveness of risk reduction, a paradigm that has many aspects. It is often called cost-benefit analysis in a narrow sense, since the benefit considered is that of risk reduction. Various actions to reduce risk may be ordered on the basis of the ratio between the magnitude of risk reduced and the magnitude of the cost of risk reduction. When smoothed, the resultant curve is concave upward (Figure 8). The curve reveals that the problem of assigning risk has

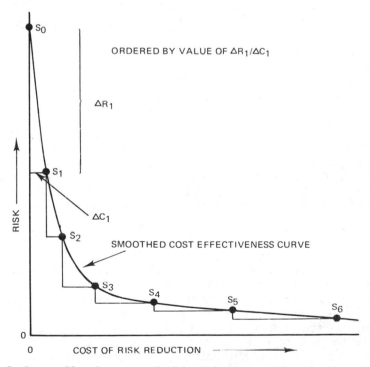

ORDERED BY VALUE OF $\Delta R_1/\Delta C_1$

ΔR_1

ΔC_1

SMOOTHED COST EFFECTIVENESS CURVE

RISK

S_0
S_1
S_2
S_3
S_4
S_5
S_6

0

0   COST OF RISK REDUCTION

Fig. 8 Cost effectiveness of risk reduction ordered relationship for discrete actions S, = risk reduction action. ΔR, = change in risk for S, ΔC, = change in cost for S, (from Rowe, 1977a, with permission).

simply been transferred to a new parameter, the cost-effectiveness of risk reduction. However, both internal and external criteria still

must be used to determine the acceptable level of cost-effectiveness.
When there is no internal breakpoint, a number of arbitrary
conditions may be considered (Figure 9), all involving external

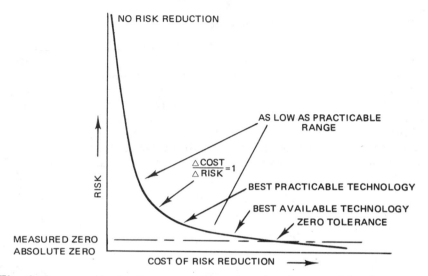

Fig. 9 Some criteria for acceptance levels of cost effectiveness of
risk reduction (from Rowe, 1977a, with permission).

references. In all cases for cost-effectiveness of risk reduction, a
referent is required to set acceptable levels of cost-effectiveness of
risk reduction. As a result, this paradigm faces the same types of
problems encountered by risk acceptance levels except that risk is
not considered directly. Economic factors are added, since the cost-
effectiveness of risk reduction is used as the primary parameter.

Risk-Benefit Balancing

So far we have considered only direct expenditures for the purposes
of minimizing risk and other similar indirect costs or for obtaining
specific benefits that involve commitments for dollar outlays. These
expenditures are direct losses and are costs in the classical economic
sense. Indirect losses, including risks, are reduced by expenditures
resulting in direct losses. When the indirect losses are all risks, a
cost-effectiveness of risk-reduction curve is developed (Figure 8).[1]
Now to make a risk-benefit balance, one superimposes, over this
curve, a curve for achieving the cost-effectiveness of obtaining

---

[1] The term "cost-risk-benefit analysis" is often seen in the literature.
The following equivalence with gains and losses is used here: costs –
Direct losses (economic); risks – Indirect losses (economic and
otherwise); benefits – Gains, both direct and indirect.

direct and indirect gains (benefits). However, this requires a scale
different from that used for losses (indirect costs). This latter curve
is convex upward, since the steps to obtain benefits can be ordered by
ratio of gain to direct cost. Both curves appear in Figure 10.

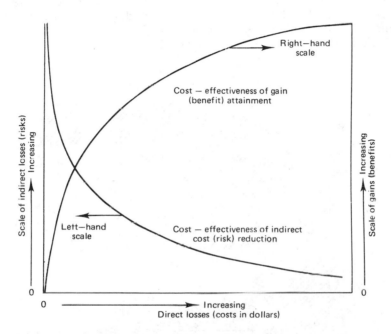

Fig. 10 Gain and indirect loss curves as a parametric function of
direct loss (cost) expenditures. (From Rowe, 1977a, with permission).

Economic theory indicates that, assuming the scales for indirect
losses and gains are identical, balancing the two curves at the margin
will provide an economically optimum condition. This means that
when the slopes of the two curves (their first derivatives) are equal,
another dollar spent to achieve benefits will be no more efficient
than a dollar spent to reduce risk.

The assumption that the two scales are identical seldom holds in
practice and is the exception rather than the rule. Attempts to find
weights to assign to the scales to equate them involves considerable
uncertainty. There are a variety of indirect losses and direct and
indirect benefits. Some losses may be equated with some benefits at
the level of individual items. The problem then is how to aggregate
the individual items to achieve an overall balance or, conversely, how
to aggregate against each scale and then make an overall balance.
This is a problem of method of aggregation.

The uncertainty in measuring each parameter is another limitation. Direct losses involve estimates that probably entail the least uncertainty. Determination of risks involves uncertainty in knowledge of exposure-risk relationships, and uncertainties in specifying intangible benefits are often much greater than uncertainties in risk estimates. Even when gain and indirect loss scales are identical, the uncertainties in measurement are so large that probably meaningful analysis is not obtained. This is illustrated in Figure 11 where the bands of uncertainty for indirect losses and gains indicate a relative basis of knowledge of each.

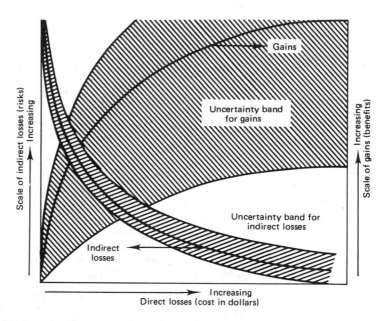

Fig. 11 Bands of uncertainty for gain and indirect loss curves. (From Rowe, 1977a, with permission).

Furthermore, the author has made an error analysis of risk-benefit analysis, assuming all required data are available (Rowe, 1977a). The conclusion is that a precision of one part in three, certainly no more than one part in ten, can result from such an analysis.

Establishing a Risk Acceptance Process

The conclusion must be that there is no single method for making risk acceptability decisions. A case-by-case examination of each decision using any combination of the above methods that work is perhaps the best that can be done. What must be recognized is that the process may be more important than the method. Thus the process, its visibility, its fairness and credibility, and its opportunity for participation may itself be the method to achieve acceptable risks under the operating definition. The process may also increase anxiety as attention is formally focused on risks formerly not perceived, but this may be a prelude to acceptability.

The outline of a formal risk assessment process that assures that all approaches are considered is described below.

A process for analysing the risk acceptance involves four distinct steps:

1) direct gain-loss analysis;
2) indirect gain-loss analysis;
3) cost-effectiveness of risk reduction; and
4) reconciliation of risk inequities (Rowe, 1977b).

First, direct gains are compared to direct losses; this analysis constitutes the classic cost-benefit analysis conducted before a new project or programme is sponsored. The individual or institution undertaking the project receives project benefits and accepts project costs, and has full responsibility for making an initial, primarily economic, cost-benefit analysis. At this stage, voluntary risks are taken to achieve specific results. If the direct gain-loss balance is negative, motivation for proceeding with the new project disappears. Unless the balance is changed or new factors, such as economic subsidization, are introduced, the project will be discontinued. A balance in favour of gain, on the other hand, will provide incentive for the programme. Public opinion, and institutional factors such as legal constraints, tax incentives or disincentives, are not always quantifiable at this stage. Direct gain-loss analysis is open-ended, because additional direct costs from subsequent steps may affect the gain-loss balance, and new factors that may change the balance must be evaluated as they occur. The sponsor, as the recipient of direct gains and losses, will continually review his position from the project's inception through its completion, if only for economic reasons.

The second step in determining risk acceptance is analysing the indirect gains and losses of risk. The indirect societal gains of a proposed activity are balanced against its indirect losses; risks themselves constitute one aspect of the societal losses. This indirect gain-loss analysis is the type of overall cost-benefit analysis sought in environmental impact statements under NEPA[2] and is a goal of most technology assessment activities.[3] Such a balance must be made, at

---

[2] 42 U.S.C. § 4332(2)(C)(1970).
[3] See Technology Assessment Act of 1972, 2 U.S.C. § § 471-481 (Supp. V 1975).

a minimum, for local, national, and world levels of impact. Because of difficulty in quantifying indirect gains and losses, this type analysis usually includes qualitative value judgements rather than numerical estimates. At the government level, a sponsoring agency is usually responsible for preparation of such analyses.

The third step in evaluating risk acceptance is determining the cost-effectiveness of risk reduction. Because society is generally risk-aversive, the risk in obtaining a particular gain or benefit must be minimized to the extent feasible, even where indirect gain-loss balances favour gain. Thus the costs of risk reduction must be computed in the analysis of both direct and indirect gains and losses. The central question in this risk reduction analysis is determining the point at which risk has been sufficiently reduced.

In an indirect gain-loss analysis, the practicality of risk reduction is one limiting consideration. The "as-low-as-practicable" limit can be defined as the point at which the incremental cost-per-risk-averted is such that a very large expenditure must be made in return for a relatively small decrease in risk, compared to the cost of previous risk reduction efforts. Thus the cost-effectiveness of risk reduction is relative, depending on the particular risk in question.

A determination of the "as-low-as-practicable" limit on risk reduction is as arbitrary as a determination of the level of risk acceptance. Some other more stable reference is required to determine when risk reduction is cost-effective or when risk decreases to an acceptable level. Thus the development and use of such a reference, based on acceptable levels of inequitably imposed risk, is the heart of an effective methodology of risk evaluation.

The fourth and most important step in evaluating risk acceptance is reconciliation of identified risk inequities. Even when an indirect gain-loss analysis is favourable, various inequities may still be imposed on specific groups in society. Such inequities occur when those who assume risks fail to receive benefits, or when risks are unevenly distributed among recipients of their benefits. In such circumstances, a risk that is otherwise acceptable because its gains outweigh its losses may become unacceptable because it is imposed inequitably. To determine acceptable levels of inequitably imposed risk, the risk in question must be identified and the type of risk ascertained.

The inequities in risk imposition can be measured against references obtained from revealed, expressed, or implied preferences as described previously.[4]   In sum, the process for risk evaluation

---

[4] Another method of establishing a risk reference is balancing risks to determine *net* risk. The risks of a new programme, including its indirect costs, are weighed against its indirect benefits to obtain net risk. The life-extending capabilities of radiation therapy in medical treatment of cancer, for example, and the increased risk of cancer

proposed here reconciles risk inequities, measured against acceptable levels of risk in the form of risk references or other external criteria, after the initial three steps in risk evaluation are performed. The process is illustrated graphically in Figure 12.

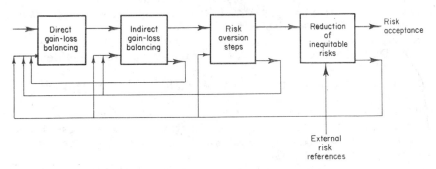

Fig. 12 Risk evaluation process. (From Rowe, 1977b, with permission).

The International Commission on Radiological Protection (ICRP) in their publication No. 26 have proposed a methodology directly analogous to the one proposed here. Balancing direct gains and losses is comparable with "justification" by an individual; balancing indirect gains and losses is comparable with societal "justification", cost-effectiveness of risk reduction is comparable with "optimization"; and limits on inequitable risks are comparable to "dose limits." While there are differences in application, the objective of both is to provide a process for setting acceptable risk levels, not of establishing the levels themselves.

The process of risk-regulation does not guarantee "right" decisions, only decisions made openly in a credible manner that assure all interested parties an opportunity to present their views and that all issues receive fair consideration. If governmental decisions are made openly, those who remain subject to inequitable risks are more likely to understand and accept the predominance of societal interests over their own. If such a process produces regulatory

---

induced by radiation therapy can be balanced. The resulting net risk is the risk to be evaluated in deciding whether to use radiation therapy. If the net risk is negative or zero, that is, if the risks balance each other, then no risk inequity exists, and a risk acceptance comparison is unnecessary. Risks can only be balanced, however, if they are measured on the same scale. This is frequently impossible; risk balancing therefore has limited application and should be used only when appropriate.

decisions acceptable to a substantial majority of our society, it will
fulfil successfully the expectations of democracy.

References

Arrow, K.J. (1963). "Social Choice and Individual Values" (Second
    edition). Yale University, New Haven, Connecticut.
Bion, W.R. (1952). Group dynamics: a re-review. *International
    Journal of Psychoanalysis 33,* 235-247.
Coombs, C.H. (1954). Social choice and strength of preference.
    In "Decision Processes" (R.M. Thrall *et al.,* eds.). Wiley, New
    York.
Council for Science and Society (1977). "The Acceptability of
    Risk". Crook-Helm, London.
Farris, D.R. and Sage, A.P. (1975). Introduction and survey of
    group decision making with applications to worth assessment.
    IEEE Transactions of Systems, Man and Cybernetics Vol.
    SMC-5, No. 3, 346-358.
Kates, Robert W. (ed.) (1977). "Managing Technological Hazard:
    Research Needs and Opportunities. Programme on Technology,
    Environment and Man, Monograph 25". Institute of Behavioural
    Science, University of Colorado.
Lowrance, W.W. (1976). "Of Acceptable Risk: Science and the
    Determination of Safety". Kaufman, Los Altos, California.
Rowe, W.D. (1977a). "An Anatomy of Risk". Wiley, New York.
Rowe, W.D. (1977b). Governmental regulations of societal risks.
    George Washington Law Review 45, No. 5, 944-968.
Warfield, J.N. (1974). "Structuring Complex Systems". Battelle
    Memorial Institute.

SUBJECT INDEX